AF839780

Costs and Benefits of Complete Water Treatment Plant Automation

Awwa Research Foundation

About the Awwa Research Foundation

The Awwa Research Foundation (AwwaRF) is a member-supported, international, nonprofit organization that sponsors research to enable water utilities, public health agencies, and other professionals to provide safe and affordable drinking water to consumers.

The Foundation's mission is to advance the science of water to improve the quality of life. To achieve this mission, the Foundation sponsors studies on all aspects of drinking water, including supply and resources, treatment, monitoring and analysis, distribution, management, and health effects. Funding for research is provided primarily by subscription payments from approximately 1,000 utilities, consulting firms, and manufacturers in North America and abroad. Additional funding comes from collaborative partnerships with other national and international organizations, allowing for resources to be leveraged, expertise to be shared, and broad-based knowledge to be developed and disseminated. Government funding serves as a third source of research dollars.

From its headquarters in Denver, Colorado, the Foundation's staff directs and supports the efforts of more than 800 volunteers who serve on the board of trustees and various committees. These volunteers represent many facets of the water industry, and contribute their expertise to select and monitor research studies that benefit the entire drinking water community.

The results of research are disseminated through a number of channels, including reports, the Web site, conferences, and periodicals.

For subscribers, the Foundation serves as a cooperative program in which water suppliers unite to pool their resources. By applying Foundation research findings, these water suppliers can save substantial costs and stay on the leading edge of drinking water science and technology. Since its inception, AwwaRF has supplied the water community with more than $300 million in applied research.

More information about the Foundation and how to become a subscriber is available on the Web at **www.awwarf.org.**

Costs and Benefits of Complete Water Treatment Plant Automation

Prepared by:
David Roberts and **David Kubel**
Black & Veatch, Kansas City, MO 64114

Alan Carrie and **Dean Schoeder**
Westin Engineering, Inc., Rancho Cordova, CA 95670
and
Chris Sorensen
Transdyn Controls, Inc., Pleasanton, CA 94588

Jointly sponsored by:
Awwa Research Foundation
6666 West Quincy Avenue, Denver, CO 80235-3098
and
U.S Environmental Protection Agency
Washington, DC

Published by:

Awwa Research Foundation

Distributed by:

American Water Works Association

IWA Publishing

DISCLAIMER

This study was jointly funded by the Awwa Research Foundation (AwwaRF) and the U.S. Environmental Protection Agency (USEPA) under Cooperative Agreement No. CR-83110401. AwwaRF and USEPA assume no responsibility for the content of the research study reported in this publication or for the opinions or statements of fact expressed in the report. The mention of trade names for commercial products does not represent or imply the approval or endorsement of either AwwaRF or USEPA. This report is presented solely for informational purposes.

TABLE OF CONTENTS

TABLES

FIGURES

FOREWORD

The Awwa Research Foundation is a nonprofit corporation dedicated to implementing research efforts to help utilities respond to regulatory requirements and traditional high-priority concerns of the water industry. The research agenda is developed through a process of consultation with subscribers and drinking water professionals. Under the umbrella of a Strategic Research Plan, the Research Advisory Council prioritizes the suggested projects based upon current and future needs, applicability, and past work. The recommendations are forwarded to the Board of Trustees for final review and selection. The foundation also sponsors research projects through the unsolicited proposal process; the Collaborative Research, Research Applications, and Tailored Collaboration programs; and various joint research efforts with organizations such as the U.S. Environmental Protection Agency, the U.S. Bureau of Reclamation, and the Association of California Water Agencies.

This publication is a result of one of these sponsored studies, and it is hoped that its findings will be applied in communities throughout the world. The following report serves not only as a means of communicating the results of the water industry's centralized research program but also as a tool to enlist the further support of the nonmember utilities and individuals.

Projects are managed closely from their inception to the final report by the Foundation's staff and a large cadre of volunteers who willingly contribute their time and expertise. The Foundation serves a planning and management function and awards contracts to other institutions such as water utilities, universities, and engineering firms. The funding for this research comes primarily from the Subscription Program, through which water utilities subscribe to the research program and make an annual payment proportionate to the volume of water they deliver. Consultants and manufacturers subscribe based on their annual billings. The program offers a cost-effective and fair method for funding research in the public interest.

A broad spectrum of water supply issues is addressed by the Foundation's research agenda: resources, treatment and operations, distribution and storage, water quality and analysis, toxicology, economics, and management. The ultimate purpose of the coordinated effort is to assist water suppliers in providing the highest possible quality of water economically and reliably. The true benefits are realized when the results are implemented at the utility level. The foundation's trustees are pleased to offer this publication as a contribution toward that end.

David E. Rager
Chair, Board of Trustees
Awwa Research Foundation

Robert C. Renner, P.E.
Executive Director
Awwa Research Foundation

ACKNOWLEDGMENTS

A research project of this nature requires support from many people in order to be successful. The input from utility participants was a key element in making sure this research is relevant and useful to AwwaRF participant needs.

The authors of this report gratefully acknowledge the participation and funding from the following organizations and individuals:

Medford Water Commission, Medford, Ore., Jim Stockton and Larry Rains
Placer County Water Agency, Auburn, Calif., Wally Cable, Brian Martin and Brent Smith
Arizona - American Water, Anthem, Ariz., Michael Helton and Dave Reves
City of Henderson, Henderson Nev., Michael Neher and Michael Morine
Onondaga County Water Authority, Syracuse, New York, Nicholas Kochan
Irvine Ranch Water District, Irvine, Calif., Carl Spangenberg
City of Austin Water and Wastewater Utility, Austin, Texas, Gary Quick
Cucamonga Valley Water Agency, Rancho Cucamonga, Calif., Ed Diggs
Northern Kentucky Water District, Fort Thomas, Kentucky, Bill Wulfeck

The authors wish to acknowledge the assistance of Julie Inman who led the literature research portion of the project and Liia Hakk for her technical editing of the report.

The advice and help of the Awwa Research Foundation project manager, Susan Turnquist, Ph.D. and the Project Advisory Committee (Nilaksh Kothari, Doug Jameson and Ramesh Kashinkunti,) are especially noted, with thanks and appreciation for their guidance on this project and commitment to the water industry - and the help of initial AwwaRF project managers Jason Allen and India Williams is appreciated.

EXECUTIVE SUMMARY

Historically, automation of water treatment plants has been justified for strategic rather than economic reasons, and usually as part of a larger project. This justification includes supporting the utility's obligation and mission to provide high-quality water service to its customers, with the cost being sometimes a secondary consideration.

A growing trend, however, is for utility management to use automation as a strategy to improve the utility's efficiency to better match the competitiveness of private industry. This approach demands a credible cost-benefit analysis. How much does automation cost? What are the added benefits? Are there risks and regulatory constraints? Will the project pay for itself? If so, how long will it take? These are typical management concerns. Private industry responds to these concerns by developing a project "business case" which includes the following components:

- The "needs" that the project will address
- The project goals and scope
- An analysis of the economic and strategic benefits
- Project costs
- Project risk

A thorough business case enables management to make an informed go/no-go decision about a proposed project, taking into account all the relevant costs, benefits, and risks. The process of developing a formal business case also helps staff to see the project in terms of its economic and strategic benefits rather than just the engineering and operational challenges.

To provide water utility decision-makers with the means to evaluate investments in automation, AwwaRF and the USEPA, sponsored this research on the costs and benefits of complete water treatment plant automation. Complete automation is defined as a level of automation that enables routine operation of the plant without on-site operators, although on-duty staff may regularly visit the plant. The definition of "Unattended" operation is no operators are on the treatment plant site for one or more shifts.

STUDY OBJECTIVES

The study had the following objectives:

- Identify the levels of automation needed for unattended operation.
- Review regulatory requirements related to unattended operation.
- Assist in identifying the benefits, risks and barriers to unattended automation.
- Develop an economic analysis method for evaluating the life-cycle cost/benefit of automation investments.
- Develop automation case studies, focused on unattended operation of water treatment facilities.

STUDY APPROACH

The study included questionnaires, a review of literature and applicable regulations, an evaluation of current economic analysis techniques, industry best practices, and case studies, to arrive at the recommendations presented in this report. A Water Utility Focus Group provided guidance during the project. The intent of this project was not to perform a statistically representative survey of the water industry regarding this topic but to provide several utility automation experiences for consideration.

Literature Research and Review

The American Water Works Association (AWWA), the Instrumentation, Systems and Automation Society (ISA), the EPA, and Water Engineering magazine are all major sources for literature on automation in the water industry. An extensive review was made of these publications looking for examples of unattended plant operation, the degree of automation used and the associated costs, benefits, and risks. The search extended beyond the water industry, to power and petrochemical industries, in an effort to learn about their experiences with unattended plant operations.

Regulatory Review

As a part of the study, federal, state and local regulations governing automation, monitoring and unattended operation of water treatment plants were reviewed.

Economic and Benefit Analysis

The methods of economic analysis evaluated included Net Present Value, Return on Investment and payback period. The NPV method is attractive because it is simple yet effective in measuring economic return and for comparison of alternatives. Combined with an evaluation of "tangible" and "intangible" benefits, it is particularly well suited for evaluating water utility automation projects. Intangible benefits are defined as benefits to which it is difficult to assign a dollar value, such as improvement of water quality, more rapid response to customer queries, or enhanced data collection. In this report, this approach referred to as the "Balanced Approach" uses many of the concepts of the highly regarded "Balanced Scorecard" method.

Development of Cost Database

An essential step in the economic analysis of a project is the development of a budgetary or "planning level" cost estimate. To assist with cost estimate development, the report includes appropriate guidelines and a reference cost database.

Risk and Barrier Assessment

Chapter 4 presents findings on some of the potential risks and barriers associated with unattended operation. Input for this chapter included responses to questionnaires completed by project participants and from available literature.

Economic Model Examples and Development of Case Studies

Six case studies were conducted with participating utilities, with focus on unattended plant operation. Five of the case studies involved unattended treatment plants. The sixth plant used a high level of automation that could support unattended operations, but the utility chose to operate it attended. The reasons for this decision are outlined in the case study summary. Appendix A includes a theoretical example of how the economic analysis method can be used for project justification.

STUDY RESULTS AND CONCLUSIONS

A major conclusion drawn from the research was that water utilities should employ recognized industry methodologies for justifying automation projects. A formal approach has been conspicuously lacking in the past. Developing a credible business case helps clarify project goals and scope and enables management to make informed decisions. The methodologies and tools provided as part of this report should help utility staff meet this goal. The following summarizes the study results and conclusions:

Literature Review

The literature review disclosed a significant body of knowledge about planning, design and implementation of automation systems for water treatment plants. A small portion of the documents reviewed also discussed unattended operation. The following is a summary of the major findings:

1. Automation is well established in the water treatment industry, and in general, operates reliably. However, better instrumentation, such as streaming current detectors, and remote notification systems would help alleviate concerns about unattended operation.
2. Limitations of automation and instrumentation were noted that make some utilities hesitant to operate their treatment plants unattended. Examples include large swings in raw water quality that make it difficult to control coagulation with simple controls. Operators often feel the need to intervene to maintain the targeted water quality parameters. These challenges can be overcome by using more sophisticated control strategies and instrumentation.
3. Utilities do not apply a consistent methodology for cost-benefit analysis of automation projects. This can make it difficult to make direct comparisons between different projects or case studies.
4. Specific data on facility performance, cost, and benefits needed for an economic analysis are often not available or are difficult to find.
5. Examples of formal justification of automation based on economics were hard to find. Justifications found, were based mostly on strategic reasons or a qualitative sense that automation would bring savings or improvements to operations.
6. Unattended plant operation correlates well with plant size. Most small surface water treatment plants are operated unattended while large plants, over 100 mgd, are continuously attended.

7. Some treatment processes such as membrane filtration require a high level of automated monitoring and control. These processes lend themselves well to unattended operations.

Regulations

Regulations pertaining to unattended operation of water treatment plants vary at both the State and local level. There are different requirements related to plant staffing, staff qualifications; and whether or not operators are required to be physically located at the treatment plant.

Some agencies allow unattended operation if the utility can demonstrate that mode of operation is successful; other agencies base their requirements on water quality and similar criteria. Several regulatory agencies simply do not permit unattended plant operation.

Federal regulations require a qualified operator to respond to an operating problem in a plant within 30 minutes. To meet this requirement during unattended periods, plants usually have one or more "on-call" operators, who respond to alarms transmitted by the plant's SCADA system.

Economic Analysis and a "Balanced" Approach

Although life-cycle economic analysis techniques are well established and widely used for water projects, the literature search found no cost-benefit analysis approach that considered both tangible and intangible benefits. However, there is a growing trend in the water industry to adopt a more comprehensive approach to evaluating investments and managing assets. The GAO asset management approach combines both life cycle cost analysis with risk analysis.

It can be difficult to justify every automation project based solely on the return on investment (ROI), that is, the "tangible" benefits. Adopting a more comprehensive "balanced" approach which considers both "tangible" and "intangible" (strategic) benefits is not only more helpful in justifying an automation project, but also more realistic.

In practice, the intangible benefits can be the major driving force. For example, the need to consistently produce high quality water or making historical data readily available to the staff for decision making are important objectives. It is difficult to assign a monetary value but these results can be key benefits from automation. The economic analysis methodology recommended in this report is therefore uses a "balanced" approach.

Another finding was that the level of automation that enables unattended operation can provide opportunities to shift production to off-peak periods to save energy costs.

Costs

The information gathered through the literature review included USEPA data that summarized the costs of new and upgraded SCADA systems, however these data did not include average or typical costs. This report provides a detailed approach to estimating budgetary costs of WTP automation and SCADA systems. This approach should be useful in conducting an economic analysis of this type of investment.

Economic Benefits

Industry data indicate that highest O&M costs at a water treatment plant are for labor, energy and chemicals. Therefore, automation in these areas has the greatest potential for producing savings. Reliably predicting savings can be challenging. This report recommends a method of estimating savings as a percentage of current (pre-automation) costs. Costs can be obtained from historical data or data especially collected for the project. The percentage savings used in the estimate should be based on published information on achieved savings for similar plants and levels of automation. If such data are not available an estimate, agreed to by operation and management staff, may be used.

An investigation of typical savings produced by applying advanced automation showed the following range of values:

- Chemical savings: Typically 15 to 40 percent
- Labor savings: Typically 5 to 30 percent, some higher values reported with unattended operation
- Energy savings: Typically 5 to 35 percent

Some of these savings may be attributable to applying a greater level of automation. Not all these savings are attributable exclusively to unattended operation.

Risks and Barriers

Chapter 4 of this report discusses the risks to be considered and mitigated when implementing automation and unattended operation at a WTP. It is notable that several utilities do not appear to consider reliability of automation a major determining factor in the decision to utilize advanced automation. Field devices such as pumps, valves and field instruments seemed to fail most frequently, since these devices are exposed to the harshest conditions. The recommended strategy for mitigating the risk of failure is as follows:

- Selecting the appropriate device during design. An appropriate device is one with proven performance in the intended environment.
- Providing regular maintenance.
- Providing on-line monitoring of the condition of the devices in the form of warning alarms for vibration, high and low tank levels, high and low residual levels, etc.

Two major reasons for not implementing unattended plant operation were reported. The first was regulatory. Several utilities indicated that state regulations prevent them from operating their plants unattended. The second was risk reduction. This reason was noted by utilities that operate large plants serving as the primary source of a community's drinking water. Management perceived unattended operation as decreasing safety and therefore compromising public health.

RECOMMENDATIONS

The following are recommendations for water utilities that are considering the costs and benefits of automation to support unattended plant operation:

1. Investigate all regulations and identify any regulatory constraints on unattended operation.
2. Carefully define the scope and goals of the automation project.
3. Evaluate the risks and consequences associated with the potential failures of automation.
4. Provide a safety margin between the operational and process goals and the regulatory limits on plant operation.
5. Develop a cost model including the capital and operating costs of automation. Do not underestimate the construction costs and the ongoing operations and maintenance costs.
6. Define both the tangible and intangible benefits of automation through brain-storming sessions with operation and maintenance staff. Quantify the tangible benefits and rate the importance of the intangible benefits. Use conservative estimates of expected savings.
7. Build consensus and management involvement early in the development of a business case for automation.
8. Develop a project business case that can be presented to management. Include both a benefit and a risk analysis. Recognize that automation improvements may be difficult to justify based solely on tangible benefits.
9. Design an automation system to support unattended operation.
10. Employ industry best practices for engineering, contracting for services, and procurement.
11. Establish a method or means to better collect historical data on plant production, energy utilization, chemical costs, and labor costs prior to completing the economic analysis.

FUTURE RESEARCH

The decision to operate water treatment plants in an unattended manner is a complex one involving more issues than economics alone. The research team encountered many cases where the financial benefits were not the deciding factors in the decision whether to operate unattended. In some cases where there was a desire to perform an economic analysis, the data was not available to support a thorough evaluation. In another case, although the utility had adequate automation to support unattended operation, due to regulations they did not operate in that mode. To address some of these overarching concerns, the following future research is recommended:

- Develop information or methods for better communication to financial decisions makers and regulators that complete automation can be a good thing. This may come in the form of a communications project.
- To assist water utilities in performing an economic analysis of their situation, it would be useful to develop a framework for economic and performance data collection. The goal would be to develop approaches that utilities can take to structure data gathering, historical data storage and performance metrics so that performance evaluation can be done on an ongoing basis. This information would allow utilities to better assess potential savings from complete automation.

CHAPTER 1
INTRODUCTION AND BACKGROUND

INTRODUCTION

As automation technologies advance and become more reliable, they are increasingly an integral part of a utility's operating strategy and facilitate unattended water treatment plant operation. The increased use of automation also makes it more common for the automation elements to represent an increasingly significant portion of capital project costs in terms of both time and money. This report outlines methods for utility decisions makers to use in analyzing the costs, benefits, and risks of automation in support of unattended plant operation.

OBJECTIVES

To assist water utility decision makers considering automation of their systems, AwwaRF and the USEPA sponsored this research to evaluate the costs and benefits of water treatment plant automation. The focus of this research report is on complete automation of water treatment plants, that is, the plant normally operates without any operators present, although personnel may make regular visits throughout the day. The definition of "unattended" operation includes no operators on-site during one or more shifts.

During unattended operation, there is usually at least one operator available "on-call". These operators typically rely on the plant Supervisory Control and Data Acquisition (SCADA) system to indicate any abnormal operating conditions and to provide off-site alarm/indication.

This report presents the results of investigations into unattended water treatment plant operation and provides an approach to economic analysis of tangible and intangible costs and benefits of automation; identification of potential risks and mitigation measures; and development of a business case for automation projects, illustrated by case studies and example evaluations.

The information is intended to be used as an aid to decision-making and to stimulate discussions during the planning of automation projects. It is not intended to be used as a detailed design guide, but rather as a part of the overall decision-making process, coupled with the appropriate utility specific considerations and engineering judgment.

REPORT ORGANIZATION

Chapter 1 – Introduction and Background

This chapter presents an overview of the research, a summary of the elements of the research, the need for economic analysis and approaches to estimating costs and benefits. It also describes the elements of a typical life-cycle cost analysis, introduces the "Balanced Approach" approach, and describes the results of the literature search.

Chapter 2 – WTP Automation Regulations and Industry Practices

This chapter presents a review of the federal, state and local regulations applicable to automation and staffing requirements for a typical water treatment plant. A summary table provides an overview of the automation regulations and plant staffing requirements for eight of the largest states (by population) in the United States.

Chapter 3 – Costs and Benefits of WTP Automation Systems

This chapter describes cost and benefit categories; provides an approach for estimating probable construction costs; offers suggestions on where to look for tangible and intangible benefits; and provides supporting information on construction cost estimating. Sample costing spreadsheets are provided in Appendix C.

Chapter 4 – Automation Considerations

Chapter 4 discusses areas of potential of risk associated with plant automation and presents recommendations on risk evaluation and mitigation measures. Minimum recommended plant wide control system design features are also presented.

Chapter 5 – WTP Unit Process Considerations

This chapter provides an overview of the types of unit processes commonly used in water treatment plants, discusses process specific automation, and outlines the degree of automation generally required for unattended operations together with representative costs, benefits, and the associated risks.

The intent is not to provide comprehensive descriptions of all possible water treatment processes but rather, how to identify the costs, benefits and potentials risks associated with process automation. The chapter also includes industry data on the savings in energy, labor and chemical costs that may be gained by implementing automation.

Chapter 6 – "Balanced Approach" Methodology

This chapter summarizes the concepts discussed in the preceding chapters and presents a step-by-step method for performing an in-depth analysis of both economic and intangible aspects of automation. This method, referred to as a "Balanced Approach," incorporates the basic elements of a traditional Net Present Value (NPV) analysis with the concepts of a *Balanced Scorecard* approach that considers the intangible benefits. A hypothetical case study, for the Rexfordingham utility, is included to demonstrate the methodology.

Appendix A – NPV Examples

Example spreadsheets are provided to demonstrate the approach to completing the NPV calculations.

Appendix B - Case Studies

Case studies, related to unattended plant operation, were conducted with participating utilities. Four of the case studies involved treatment plants that operate unattended. One case study involved a plant that has a high level of automation and could operate unattended, but the utility has chosen not to operate the plant in this mode. The reasons for this decision are presented in the case study summary.

The case studies used elements of the Balanced Approach; however there is significant difference in the level of detail in the various case studies, primarily due to the level of information available at the time of the analyses. The primary value of the case studies is to stimulate thought and discussion on various scenarios related to automation.

Appendix C – Cost Database and Example Cost Estimate

A cost database is included with unit pricing information to be used to develop planning level cost estimates for WTP automation projects. An example cost estimate is included.

Appendix D – Literature Review and Search

The results of the literature search are presented in Appendix D.

DRIVERS OF UNATTENDED WTP OPERATION

Water utilities face a variety of changes and trends that impact their operations, maintenance, and capital expenditures including the following:

- Deteriorating quality and declining quantity of water supplies
- Increasing regulatory and reporting requirements
- Increasing need for adding and replacing infrastructure
- Advances in water treatment technologies
- Increasing resistance to higher water rates and potential for financial crisis
- Consumer expectations for higher quality water at lower costs
- Utility consolidation, reducing the number of small utilities
- Shortage of skilled workers
- Increasing energy costs
- Increasing chemical costs
- Increasing labor costs

Automation can help utilities mitigate and alleviate the impacts of many of these changes. Automation that enables unattended plant operation can have a significant impact on several of these fronts.

REGULATIONS AND UNATTENDED PLANT OPERATION

While automation can eliminate many of the technological barriers to unattended WTP operation, many utilities do not operate in this mode for a variety of reasons. These reasons include regulatory requirements, economic considerations and concerns over treated water quality.

Federal, state and local drinking water regulations influence the treatment decisions, especially those pertaining to unattended plant operation. Current, pending and anticipated future regulations have a direct or indirect impact on the types of instrumentation and monitoring, reporting and automation practices used at water treatment facilities. Examples of these regulations include:

- Long Term 1 Enhanced Surface Water Treatment Rule (LT1ESWTR)
- Long Term 2 Enhanced Surface Water Treatment Rule (LT2ESWTR)
- Stage 2 Disinfectants/Disinfection By-Products Rule (Stage 2 DBPR)
- USEPA Small Systems Requirements
- Water System Security Legislation, Vulnerability Assessments, and Distribution System Monitoring Regulations

This report describes regulatory considerations and many of the risks and barriers to unattended water treatment plant operation.

DRIVERS OF ECONOMIC ANALYSIS

Historically, economic analysis for automation projects included preparation of a construction cost estimate, with little focus on developing a business case for the expenditures. Where a business case was required, expenditures for automation were typically considered a minor part of the overall cost/benefit assessment of a capital improvement project. Automation, where used, was justified on the basis of its necessity, or benefits to the overall capital improvement program. As the use of automation has become more prevalent and its benefits to utilities are more widely recognized, large stand-alone automation projects have become more common. Consequently, there is a growing need to develop detailed business cases for automation projects.

Although automation depends on reliable technology, in the form of computers, application software, networks, communications and field instrumentation, this technology should be viewed as a means of supporting the automation and business goals, not as an end in itself. With automation and operating strategies becoming more complex, the utility manager needs to balance a large number of sometimes, conflicting requirements.

Considerations include the risks inherent in unattended operation, economic constraints, security, customer support, staffing, and regulatory requirements. With increasing pressure on utilities to operate more effectively, managers need information and methodologies to help them make the decisions. This research effort has confirmed that the water industry has no standard approach or guidelines for economic analyses of automation that includes the development of business cases.

In private, or investor owned business enterprises, automation can be and is justified based on ROI, because a return is expected and measured. Investments in automation can increase production as well as reduce the costs of production, generating both more revenue and a higher profit margin. This is not the case with non-profit public agencies. Automation has the potential to reduce operation and maintenance reduces costs, but generally does not increase revenues. There is no profit "return" to measure, no competitive leverage to drive growth.

Many public utilities use the Net Present Value (NPV) based life-cycle cost analysis for capital improvement projects. In NPV analysis, the costs and benefits of a project are expressed as an equivalent cost in today's dollars. This method can be used in comparing different alternatives that may have different cash flow profiles throughout the expected life-cycle. This technique makes it possible to compare projects with lower initial costs and higher annual expenses with those projects that have a higher initial cost but lower recurring costs.

Although ROI and NPV analyses are appropriate for many situations, they usually do not consider benefits that are more difficult to quantify, such as greater reliability and emergency response capabilities; avoided costs as a result of better maintenance, improved operation, process improvements, and better regulatory compliance.

The majority of intangible benefits that drive automation related decisions in the public sector are various forms of risk mitigation. Automation can reduce risk of adverse consequences of poor water quality, personnel availability, service outages or low pressure, taste and odor episodes, security breaches, and others.

The need for a rigorous economic analysis for automation projects was the major driver behind this research and was identified in a previous research project as an industry wide need. The need to justify automation related expenditures was also identified by several of the participating utilities as an important element of the overall automation decision process.

UNDERSTANDING THE COSTS AND BENEFITS

The costs and benefits of automation projects need to be understood as a part of the overall decision to authorize a project. These costs and benefits can be tangible (objective and quantifiable) or intangible (subjective and unquantifiable).

Tangible costs of automation projects typically focus on engineering and construction costs. Other quantifiable costs that should be considered but are frequently overlooked include software and hardware maintenance, future upgrading, and staff training. Sources available for estimating costs include construction cost estimating manuals, vendor information, and industry benchmark data.

Intangible costs can include the disruptive effects of organizational and procedural changes associated with introducing a new technology and the effort required to overcome regulatory or personnel concerns.

Tangible benefits of automation can include reduction in labor cost; ability to add processes or to support plant expansion without adding staff; reduction in travel to remote facilities; lower chemical costs as a result of better dosage control, and reduced energy costs as a result of process optimization and/or off-peak pumping.

Intangible benefits can include items to which it is difficult to assign an economic value, such as improved finished water quality, automated regulatory reporting, improved collection and handling of historical data, improved staff morale and better documentation.

Tangible Costs

There is a variety of sources available for estimating the tangible costs of automation projects. However, due to the complexity of most control systems, and the numerous system elements that need to be estimated; estimating these can be a difficult task. A number of factors need to be considered in developing an estimate of probable cost for an automation project including: the existing facility conditions; level of documentation; condition of mechanical and process equipment; physical arrangement of the facilities; plant capacity; the number of sites; location where operators interact with the system, and the approach to procurement. Given the complexity of automation projects there is a general desire among utility engineers and managers to simplify the cost estimating and to develop rule of thumb estimating techniques. Figure 1.1 from the USEPA publication *Drinking Water Infrastructure Needs Survey, Modeling the Cost of*

Infrastructure, (1999), includes data on the cost of SCADA system rehabilitation projects for water treatment plants of various capacities.

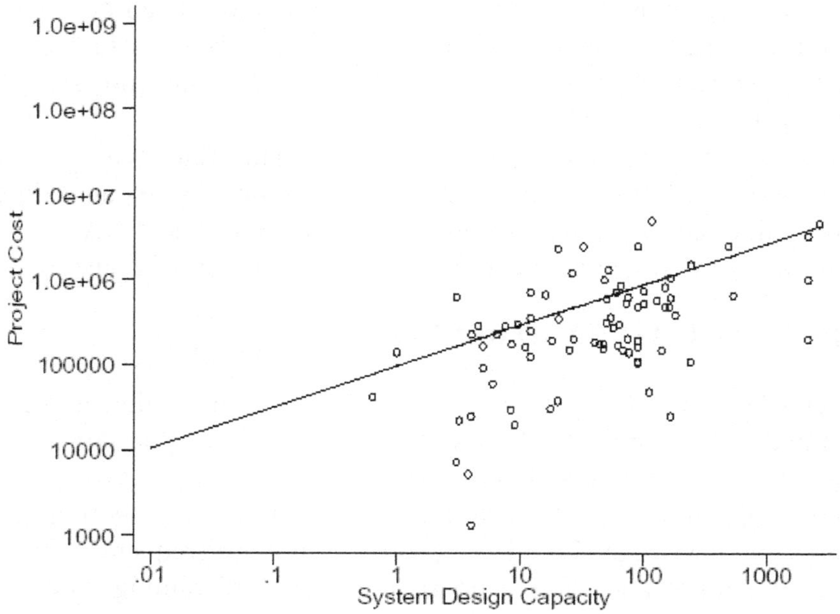

Source: USEPA 1999.
Figure 1.1 Costs of computer and automation (SCADA) system rehabilitation

Figure 1.2 shows cost data for computer and automation associated with new water treatment plants of varying capacity. These charts illustrate the wide range of encountered costs associated with automation projects for water treatment plants and highlight the difficulty of attempting to develop standardized "rule of thumb" approaches to cost estimating.

This report provides a practical project assessment approach to estimating a probable or budgetary cost of automation improvements.

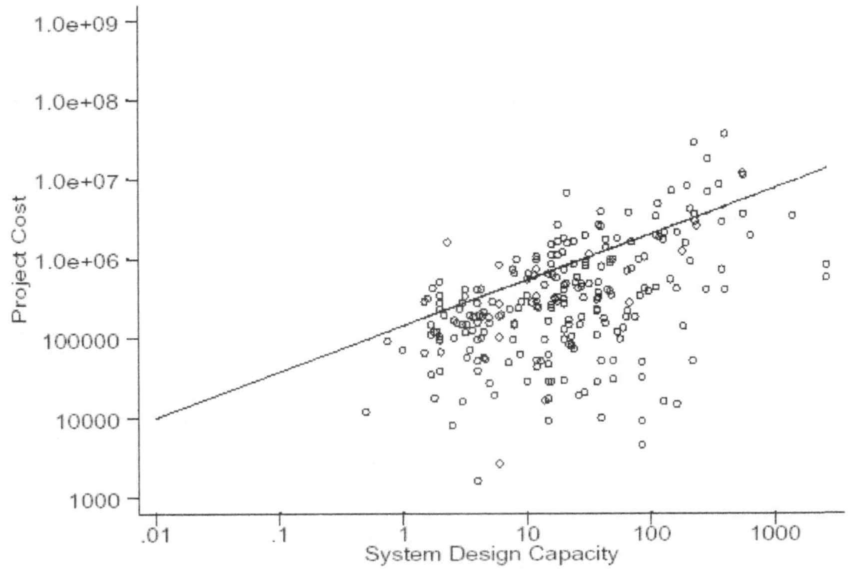

Source: USEPA 1999.

Figure 1.2 Costs of new computers and automation (SCADA) systems

ECONOMIC LIFE CYCLE COST ANALYSIS

The economic analysis of life-cycle costs is common in estimating the cost of engineering projects in the water industry. The goals of a typical life-cycle cost analysis include quantifying the tangible costs and benefits associated with planning, procurement, operation, maintenance, and ultimately disposal of project elements. Construction cost is an important component of the analysis; however, equally important is the total cost of ownership beyond the initial cost.

The cost-benefit assessment method recommended by the Federal Government for projects is outlined in "Circular No. A-94, Revised (Transmittal Memo No. 64), October 29, 1992, *Guidelines and Discount Rates for Benefit-Cost Analysis of Federal Programs*". The analysis includes the Net Present Value approach, which expresses the costs and benefits over the life of the project in terms of a net present cost or value. These costs include capital expenditures, operating costs, maintenance, training, and salvage value amortized over the life of the project. Benefits can include savings in labor, energy, and chemical costs; reduction in fines, all of which can also be expressed as a present value. Other financial considerations include the cost of money, inflation rates, life of the project, and costs of lost opportunity.

For a typical analysis, the costs and benefits of a project over time and the duration or lifecycle of the project are identified. For control system equipment, the life cycle may be 2 to 4 years or less for computers; 5 to 7 years for software and some hardware; and 15 to 20 years for instruments, control panels, and wiring.

Although NPV and ROI analyses are appropriate for many situations, they typically do not consider benefits that may be more difficult to quantify such as increased reliability, emergency response capabilities, avoided cost due to enhanced maintenance, improved operation, business process improvement, and enhanced ability to maintain regulatory compliance.

STRATEGIC COSTS AND BENEFITS

Several approaches can be used to incorporate tangible and intangible costs and benefits into the decision making process. This section covers two approaches:

1. Balanced Scorecard
2. Asset Management

Balanced Scorecard

Kaplan and Norton described an approach called "The Balanced Scorecard" in their 1996 book with the same title. The Balanced Scorecard approach begins with the organization's primary vision and mission with investment decisions divided into four categories:

- Financial Impacts – are we investing responsibly and are there tangible benefits?
- Customer Impacts – are we providing good service and how do our customers view us?
- Business Process Impacts – are we efficient and providing value?
- Learning and Growth – are we improving as an organization?

The Balanced Scorecard approach to investment decisions includes both financial and non-financial goals, and can be used by both the private sector and the public sector. It involves developing a scorecard rating for projects, assigning relative weights to strategic objectives, and providing a balanced look at how the project benefits the organization and meets the needs of customers.

The Balanced Scorecard provides a framework for making management decisions according to the needs of the specific project or issue analyzed, in the context of the overall goals of the organization. In developing an example scorecard for an automation project, the four organizational considerations listed above are further divided into the core strategic objectives for the organization, which are then prioritized by a weighting factor. The rating for a project-specific consideration is combined with the priority of the associated organizational consideration, to determine the overall rating for each. Financial impacts might be broken down and prioritized as indicated in Table 1.1. In developing the project-specific portion of the scorecard, each project specific consideration is associated with one or more strategic utility objectives, and rated according to its effect on the associated strategic consideration. Using the financial impacts as an example, a portion of a representative "scorecard" weighting could be as indicated in Table 1.2.

Table 1.1
Organizational strategic financial objectives

Consideration	Strategic Utility Objectives	Priority
Financial	Operating Expense Reductions	Med
	Optimizing Asset Use	High
	Growth in Service Area	High

The rating of the project-specific consideration is combined with the priority of the associated strategic consideration to determine its overall rating. In the example in Table 1.2, the total rating is obtained by multiplying the numeric value of the priority (Low = 1, Med = 2 and High = 3) by the project rating. In the above example, although the project does not result in significant savings in staff or energy, it is important because of its ability to support growth in the service area.

This analysis method may incorporate other areas of organizational consideration such as customer impacts, business process impacts, and learning and growth. After the overall rankings are determined, a more traditional life-cycle cost analysis is performed by combining the scorecard rankings with the tangible and intangible costs and benefits to provide a "balanced" perspective on the business value of the project. The method presented in this report is a simplified adaptation of the Balanced Scorecard approach.

Asset Management

The Government Accountability Office (GAO) has prepared a draft report (GAO-04-461) on comprehensive asset management to identify needs and to plan for future investments. The GAO forwarded the report to the USEPA for review and comment on its applicability for planning infrastructure improvements.

Asset management based principles in the water and wastewater industries is are the early stages of adoption. The GAO approach recommends consideration of the life cycle and total cost of ownership concepts and includes considerations of risk and level of service but does not provide clear guidelines for the consideration of intangibles.

LITERATURE REVIEW

A key element of this project was a literature search for relevant information on automation for water and non-water industries. Some of the findings of the literature search are discussed below. Additional information on this subject is in Appendix D. The literature search included the following topics:

- Technology Trends
- Automation, Planning, Design, Procurement, and Implementation
- Water Treatment Process Optimization

Table 1.2
Project specific financial objectives and ratings

Consideration	Strategic Utility Objectives	Priority	Indicator	Project Rating (1 low to 10 high)	TOTAL RATING
Financial	Operating Expense Reductions	Med	Reduction in plant shift staffing levels	2	4
	Operating Expense Reductions	Med	Reduction in energy costs	1	2
	Optimizing Asset Use	High	Maximizing the use of plant capacity	1	3
	Growth in Service Area	High	Existing System cannot be expanded	8	24

- Energy Management
- Cost-benefit Analysis of Automation
- Water Industry Regulations
- Non-Water Industry Automation

Detailed findings from the literature search include:

Technology Trends

Although the water industry tends to be conservative in its deployment of new technologies, it is adopting technologies such as computers, wireless communications, advanced instrumentation, control systems and automation at an increasing rate.

Means et al. (2006), in an overview of technology trends and their implications for water utilities, found that information and technology advances are finding their way into every aspect of the water industry, and bringing along greater efficiency. They also noted, "Automation of water treatment is likely to expand as new technologies require less hands-on management and water utilities press to reduce labor and operating costs."

The move toward unattended operation of water treatment plants will depend primarily on the availability of reliable technologies. The trends indicate a growing refinement and adoption of such technologies, which should further increase their use.

Automation Planning, Design, Procurement, and Implementation

The literature search turned up a significant amount of information on planning, design, procurement and implementation of automation systems for water treatment plants.

This research builds upon previous work by the water industry and research by AwwaRF into the use of automation in the treatment and distribution of drinking water. Numerous sources of information are available on automation for water treatment plants. One reference that identified the need for this research project is the 1996 AwwaRF report, *Automation Management Strategies for Water Treatment Facilities*, which provides information and perspectives of the water industry regarding automation. Some of the specific technologies have been upgraded since its publication but the report provides a base of understanding of the issues involved.

The AWWA Manual of Practice M2 and other industry reference materials contain additional background information on process automation and operating strategies for water treatment facilities.

Water Treatment Process Optimization

An understanding of water treatment processes and the automation needed for unattended operation of these processes is a key component of this research. Numerous references are available on this subject from AWWA and AwwaRF. One of the most widely used references is the AWWA 2005, Fourth Edition, *Water Treatment Plant Design*, which includes industry-accepted design practices as well as a discussion of theory, design considerations and design criteria for water treatment processes.

In addition to these references, many studies and standards are available on optimization strategies for specific unit processes including AWWA conference proceedings, standards, and manuals of practice and AwwaRF studies. An example is *An Evaluation of Streaming Current Detectors* (Dentel, Kingery, 1988), pertaining to the automation of coagulant dosing, which presents numerical results, including cost and payback periods, for ten water treatment plants that practice automatic coagulant control using a streaming current detector. AWWA 2000, *Manual of Practice M37, Operational Control of Coagulation and Filtration Processes,* describes in detail the methods used to optimize coagulation and filter processes.

Energy Management

In addition to process optimization, water treatment plants can realize significant benefits through management and optimization of energy use. Research by AwwaRF, The California Energy Commission, the EPRI Municipal Water & Wastewater Program, The American Council for an Energy Efficient Economy - Energy Efficiency in the Water and Wastewater Sectors, and the Department of Energy, into energy efficiency in water and wastewater systems, which is currently underway, is expected to lead to more thorough understanding of energy saving opportunities. Currently available reference material includes the following:

- EPRI (Electric Power Research Institute) 1996, *Water and Wastewater Industries: Characteristics and Energy Management Opportunities.*
- EPRI (Electric Power Research Institute) 1994, *Energy Audit Manual for Water/Wastewater Utilities.*
- AwwaRF/EPRI/CEC 1997, *Quality Energy Efficiency Retrofits for Water Systems.*
- EPRI (Electric Power Research Institute) 2001, *Summary Report for California Energy Commission Energy Efficiency Studies, Appendix 2.7: Water and Wastewater Treatment Plant Energy Optimization Evaluations,* Palo Alto, Ca.
- Jacobs, J. J., Kerestes, T. A., Riddle, W. F. 2003, *Best Practices for Energy Management,* AwwaRF, Denver, Colo.

Cost-Benefit Analysis

One of the objectives of this research is to develop methods of economic analysis for planning automation projects for water treatment plants. The literature search resulted in identifying a significant body of literature on methods of economic analysis used by a wide variety of industries, which include internal rate of return, net present value, return on investment and payback period, among others. This massive body of literature was condensed to documents considered most relevant to the water utility industry.

One such document is Circular No. A-94, *Guidelines and Discount Rates for Benefit-Cost Analysis of Federal Programs* (U.S. Government, 1992), which recommends the NPV economic analysis approach for infrastructure projects and provides guidelines for developing cost-benefit analysis for federal projects.

Other relevant documents include the U.S. Department of Energy, Federal Energy Management Program, publication 10 CFR 436, Subpart A, *Methodology and Procedures for Life Cycle Cost Analyses* and the U.S. Department of Commerce, NIST Handbook 135 *Life Cycle Costing Manual for the Federal Energy Management Program.* Combined, these documents serve as the basis for the NPV-based method of life cycle cost analysis methodology presented in this study.

While the NPV economic analysis addresses tangible costs and benefits, it is also important to incorporate the intangible costs and benefits to present a balanced approach case. It was found that there are fewer literature sources discussing methods of economic analysis that incorporate intangible costs and benefits; however, Kaplan and Norton present such an approach in their book *The Balanced Scorecard* (Kaplan and Norton, 1996). This approach, which considers an organization's primary goals, both financial and non-financial, was initially directed to the private sector, but has been used in the public sector as well. Many of the principles from Kaplan, Norton 1995, *The Balanced Scorecard,* were used in developing the assessment method discussed in this research project.

Water Industry Regulations

The literature search and review included federal, state, and local drinking water regulations with a focus on regulations governing operational monitoring, staffing and unattended operation of water treatment plants. To provide a representative review of government regulations for this report the research was limited to the eight largest states in terms of population: California, Florida, Illinois, Michigan, New York, Ohio, Pennsylvania, and Texas.

All federal and state regulations are available on-line through the respective agencies' websites. Review of the regulations indicates that operational testing requirements (instrumentation and data gathering) do not present a barrier to unattended operation of water treatment plants. Regarding staffing and unattended operation of water treatment plants, the U.S. EPA Community Water System Regulations (1999) mandate that each state develop an operator certification program that incorporates the following:

- Classification of community water systems based upon potential health risks
- Owners must place the direct supervision of the system under the charge of an operator holding a valid certification equal to or greater than the system classification
- A certified operator must be designated and "available" for each operating shift

The federal guidelines serve as the basis for state classification and staffing requirements. Although each state uses a slightly different approach, they all have a classification system for community water systems based on source and quality of the water supply. In general, water systems that have a consistent, high quality source have minimal certified operator and staffing requirements. The classification of some of the states is further differentiated according to capacity and/or number of people served. Smaller systems typically have lesser requirements for certified operators and plant staffing.

Non-Water Industry Automation

The literature review included information on advanced automation from non-water industries such as wastewater, fossil-fueled electric power generation, hydroelectric power, and petrochemical industries, covering lessons learned and applicability to water utilities. This research revealed that the wastewater industry is similar to the water industry in that it lacks both standardized economic analysis methods and cost data. The results of a survey of wastewater utilities by Hill et al. (2002) indicate that most of the respondents justify having installed automation systems because of the associated cost savings; however, less than 10% of the facilities surveyed had data to support this claim. Hill also recommends that further research be conducted to compare the costs and performance of a wastewater treatment plant before and after implementation of a comprehensive monitoring and control system.

Literature related to advanced automation for fossil-fueled electric power, hydroelectric power and petrochemical industries revealed that these industries employ significantly more rigorous and systematic approaches to economic analysis of automation projects. EPRI issued a report in 1989 titled *Hydropower Plant Modernization Guide, Volume 3: Automation*, which includes procedures for hydroelectric utilities to identify the most suitable and cost-effective implementation of automation and discusses the levels of automation for semi- and fully-automatic, remotely controlled, and unmanned sites.

The guide also includes methods for detailed economic evaluation using NPV. In a more recent study, Benson (2005) investigated the costs and benefits and evaluated options for automation, staffing levels, and responsibilities at six hydroelectric plants. An economic analysis of several alternatives indicated the District could realize payback in 1.9 to 4.7 years by reducing staffing levels. However all of the alternatives had various levels of risk associated with them. The study recommended that the risks be evaluated and mitigation strategies identified before selecting the automation alternative to be implemented.

SIGNIFICANCE OF THE PROJECT

Who should read this report and why? The authors believe that this report provides a unique and comprehensive source of information, methodologies and examples for use by decision makers involved in the evaluation and planning of water treatment plant automation projects; specifically automation projects that facilitate unattended operations. It strives to provide information not only on the technical aspects of automation but also from a business perspective. These objectives of this research will have been achieved if this report is practical to use and provides the following benefits to the water utility community:

- A reference source for information on the current levels of automation available, requirements of different processes, and regulations that impact automation decisions
- The findings of literature research and insights from other utilities, including wastewater, hydroelectric, fossil fuel power, and international water utilities
- Provides representative cost data for plant automation on design, capital costs, labor and maintenance costs to facilitate development of budgetary cost estimates for projects under consideration
- Utility case studies and sample economic calculations to enhance understanding the issues and concepts
- Information on typical risks and practical mitigation measures based on utilities' experience
- Tools that will allow a tailored analysis to the unique utility situations
- A "balanced" analysis approach for evaluating tangible and intangible costs, benefits and risks as they align and support the mission of the utility in serving customers
- Identification of potential barriers to implementing complete automation of water treatment plants and mitigation strategies.

SUMMARY

The costs and benefits of unattended operation of water treatment plants, and the automation necessary for this mode of operation, are multi-faceted and complex. This report attempts to provide a perspective on this topic, by combining the technical aspects of automation with the basic concepts of a typical business case, to offer utility decision-makers relevant information for evaluating whether or not to operate their facilities unattended.

CHAPTER 2
WTP MONITORING AND CONTROL REGULATIONS AND INDUSTRY PRACTICES

The licensing and operation of water treatment plants are regulated by the USEPA and state and local health departments. These regulatory agencies also provide guidelines for the monitoring, control, and staffing of the plants. This chapter presents an overview of current regulations, how they affect a utility's ability to operate unattended and what effect they may have on staffing levels. For the purposes of this report a representative sample of water treatment plants in the eight largest states, by population, in the United States was reviewed.

OVERVIEW

NPDWR (USEPA), state, and local drinking water regulations influence the treatment decisions for most water utilities, including the level of instrumentation, automation, and unattended operation. The following rules, pending regulations, and anticipated future regulations have a direct or indirect impact on the types of instrumentation, monitoring, reporting, and automation used at water treatment facilities:

- Long Term 1 Enhanced Surface Water Treatment Rule (LT1ESWTR)
- Long Term 2 Enhanced Surface Water Treatment Rule (LT2ESWTR)
- Stage 2 Disinfectants/Disinfection By-Products Rule (Stage 2 DBPR)
- USEPA Small Systems Requirements
- Water System Security Legislation, Vulnerability Assessments, Distribution System Monitoring
- USEPA Community Water System Requirements, 1999

STATE AND FEDERAL REGULATIONS GOVERNING OPERATIONAL MONITORING OF WATER TREATMENT PLANTS

The degree of treatment process monitoring varies among the eight states studied. Continuous monitoring is required only for turbidity and disinfection residual, both of which can be monitored using on-line analyzers and monitored remotely. Testing for other parameters is required less frequently and in some cases can be accomplished through periodic, direct observation. An overview of the monitoring requirements is in Table 2.1. It appears that the testing and reporting requirements, in general, do not present a regulatory barrier to unattended operation of a water treatment plant. In general, the instrumentation needed for continuous monitoring is widely available and reliable.

TABLE 2.1*
Operational monitoring requirements

Parameter	USEPA	Calif.	Fla.	Ill.	Mich.	N.Y.	Ohio	Pa.	Texas
IOC									
Ground water	1	9	18	13	18	17	17	13	13
Surface water	2	10	14	14	14	4	4	14	14
Nitrate									
Ground water	2	4	4	4	4	4	4	14	3
Surface water	3	3	3	3	3	3	3	14	2
Organics									
Initial Detection	3	3	3	3	3	3	3	3	3
No Initial Detection	4	4	4	15	4	4	12	4	4
Radionuclides									
Initial > MCL	3	11	3	16	3	3	3	3	3
Initial <MCL	5	5	12	17	19	5	19	17	19
Microbial Contaminants	6	6	6	6	6	6	6	6	6
Disinfection Residuals	7	7	7	7	7	7	7	7	7
Turbidity	8	8	8	8	8	8	8	8	8

Source: Adapted from various federal and state regulations.

* Sampling Key for Table 2.1	
1	Initially, one sample per compliance period; if a sample is > 50% of the MCL, then sample quarterly
2	Initially, one sample annually; if a sample is > 50% of the MCL, then sample quarterly
3	Quarterly sampling
4	Annual sampling
5	Sample every 4 years
6	Frequency dependent on population served
7	Continuous monitoring at entrance to distribution system
8	Continuous monitoring at combined filter effluent, record value every 15 minutes
9	Sample once per compliance period; if there is a persistent trend toward higher levels, then quarterly sampling
10	Sample annually; if there is a persistent trend toward higher levels, then quarterly sampling
11	Sampling frequency at the discretion of the State
12	Biannual sampling
13	Sample every 3 years, if exceed MCL, then sample quarterly
14	Sample annually; if exceed MCL, then sample quarterly
15	SW systems sample annually; GW systems sample every 3 years
16	Monthly sampling
17	Sample every 3 years
18	Sample once per compliance period; if exceed MCL, then sample quarterly
19	Sample every nine years

REGULATIONS GOVERNING PLANT STAFFING AND UNATTENDED OPERATION

Federal Regulations

In 1999, the U.S. Environmental Protection Agency (EPA) developed guidelines for the certification of operators of Community Water Systems (CWS). The guidelines mandate each state to develop an operator certification program that meets the following criteria:

1. All CWS must be assigned a classification based on indicators of potential health risk including system complexity, system size, source of supply, and extent of treatment facilities.
2. Owners of all CWS must place the direct supervision of the system under the charge of an operator holding a valid certification equal to or higher than the classification of the CWS.
3. A certified operator must be designated and "available" for each operating shift.
4. These criteria serve as the basis for State classification of CWS and associated staffing requirements.

State Regulations

This section presents a summary of the classification system and staffing requirements for plants in the eight most populated states in the study. It is intended to give an overview of the approach taken in applying the Federal regulations, rather than a comprehensive review of all possible approaches. The interpretation, requirements, and permit compliance are typically determined by state and local health departments.

Classification of CWS

Each of the eight states has developed a classification system for CWS under its jurisdiction. Even though each state takes a somewhat different approach, each state's classification system is based on the source of water supply (e.g. surface water, groundwater under direct influence of surface water, groundwater) and the quality of that supply. For example, a groundwater devoid of contaminants, both microbiological and chemical, will be classified as having minimal certified operator and staffing requirements.

Conversely, a surface water supply containing pathogenic microorganisms and chemical contaminants will receive a classification that requires certified operators with the highest level of qualifications and will be subject to the most stringent staffing requirements.

Some states further classify CWS on the basis of system capacity and/or the number of customers served. Generally, smaller systems are subject to lower requirements for certified operators and plant staffing.

Staffing Requirements

California

California requires a chief operator or shift operator on-site whenever a treatment facility is in operation. An exception is granted if the CWS has a plan of operations that "demonstrates an equal degree of operational oversight and reliability with either unmanned operation or operation under reduced operator certification requirements." In this case, the chief operator or shift operator is not required to be on-site but "shall be able to be contacted within one hour."

Florida

In Florida, systems with treatment plants that require the greatest degree of operator supervision (Class A, B, and C plants) must employ a full-time lead or chief operator for each treatment plant. Full-time is defined as at least 4 days per week for a total of 35 hours each week. Upon approval from the State, the lead/chief operator must be "available" whenever the plant is in operation. "Available" means able to be contacted as needed to initiate the appropriate action in a timely manner.

For Class A, B, and C plants, a certified operator "shall be on-site and in charge of each required shift and for periods of required staffing when the lead operator is not on-site." Daily staffing hours may be reduced upon written approval from the State for those plants that employ an electronic surveillance system or have an automatic control system.

Illinois

The regulations of Illinois do not specifically address reduced or unattended operation of CWS. Reference is made to the requirement that "all portions of a CWS shall be under the direct supervision of a properly certified operator." As is the case with other states, the more challenging the nature of the source of supply, the greater the qualification requirements for certified operators.

Michigan

In Michigan, treatment facilities that have an F Classification are required to employee a certified operator with the highest level of qualifications. In terms of staffing requirements, Michigan regulations specify that "a shift operator be on site and in charge of each operating shift at a community supply in the F classification when the operator in charge is not on site." The State may waive this requirement upon approval of a plan of operation submitted by the CWS that demonstrates "that public health will be adequately protected when a certified shift operator is not on site." Presuming a CWS can develop a State-approved operating plan, it may be granted permission for reduced staffing levels and/or unattended operation of treatment facilities.

New York

In the State of New York, CWS must "place direct supervision of the water system under an operator with certification level equal to or greater than that required for the classification of the treatment plant(s) serving the system." With regard to unattended operation, the regulations simply state that "a designated certified operator must be available during plant operation." The term "available" is not defined but is assumed to mean that the operator is not required to be on-site during plant operation but must be in a position to respond in a timely manner when needed at the treatment plant.

Ohio

Staffing requirements in Ohio vary from a minimum of three non-consecutive 30 minute visits per week at Class I systems (small groundwater systems) to 8 hours per day for 5 days per week for Class III and Class IV systems (surface water and large groundwater systems). The minimum staffing requirement for Class III and IV systems may be reduced to 2 hours per day for 5 days per week based on approval of an operating plan that describes the level of automation and continuous monitoring at the treatment facility. Also required is a detailed operations schedule that specifies the number of operators, the certification level of each operator, and the number of hours spent at the treatment facility.

Pennsylvania

The Pennsylvania regulations do not specifically address unattended operation of CWS. They establish a classification system for treatment plants based on the source of supply and the level of treatment provided, as well as the minimum experience level that a certified operator must possess in order to operate each class of treatment facility.

Texas

The regulations governing staffing requirements in Texas specify the minimum number of hours that a certified operator must be present at a treatment facility. The higher the classification of the treatment plant, the greater the number of hours of attended operation. The requirement ranges from employment of a licensed operator for small groundwater systems to 16 hours per month for large groundwater systems, to 32 hours per month for each of two operators for large surface water systems. No provisions are included to allow a reduction in the number of hours by a certified operator.

INDUSTRY PRACTICE

The results of a survey by the USEPA on community water systems and on the percentage of plants attended by operators around the clock versus plant production levels are in Table 2.2.

These data show that many plants, smaller than 100 mgd, operate without 7 days a week, 24 hours a day operators on site. However, only a few plants larger than 100 mgd operate without around-the-clock operators present. The data also indicates that the vast majority of all plants sized 10 mgd and larger have an operator present 100 percent of the time. For smaller systems, the data tends to indicates that they are likely to be operated in an unattended manner.

Table 2.2
Operator hours versus plant size

Treatment Plants and Operator Hours per Week
By Primary Source of Water and Average Daily Production

	Plant Average Daily Production (MGD)						
Water Source	**0 - 0.01**	**0.01 - 0.1**	**0.1 - 1.0**	**1.0 - 10.0**	**10.0 - 100.0**	**Over 100.0**	**All Sizes**
Ground Water Plants							
% of Plants with 24/7 Operator	2.5	0.2	0.7	12.7	52.0	0.0	1.7
Avg. hours/week for systems without a 24/7 Operator	3.3	6.6	18.3	28.4	20.6	8.0	10.0
Observations	106	157	303	275	49	1	891
Surface Water Plants							
% of Plants with 24/7 Operator	0.3	0.0	4.2	48.4	92.8	100.0	22.1
Avg. hours/week for systems without a 24/7 Operator	5.8	18.3	49.6	91.5	50.6	*	43.5
Observations	28	79	138	178	245	25	693
Mixed Plants							
% of Plants with 24/7 Operator	5.8	0.0	0.0	50.2	83.2	*	22.4
Avg. hours/week for systems without a 24/7 Operator	2.0	17.6	49.0	68.2	42.8	*	27.4
Observations	3	3	8	16	30	*	60
All Plants							
% of Plants with 24/7 Operator	2.5	0.1	1.2	28.7	84.3	95.7	4.6
Avg. hours/week for systems without a 24/7 Operator	3.4	7.6	23.6	48.9	33.0	8.0	13.8
Observations	137	239	449	469	324	26	1644

Source: USEPA 1999.

CHAPTER 3
COST AND BENEFIT CONSIDERATIONS OF AUTOMATION SYSTEMS

INTRODUCTION

This chapter presents a review of the major components of typical water treatment plant automation systems; the cost and benefit categories; a detailed approach to estimating planning-stage construction costs; operation and maintenance costs; evaluation of potential benefits; and strategies for minimizing life cycle costs.

QUANTIFYING THE COSTS AND BENEFITS

One of the stated goals of this project is to provide information to assist decision makers in planning and authorizing automation projects. A key to better planning is to understand the costs and benefits of automation systems for treatment plants, specifically those that can facilitate unattended operation. However, it is difficult to prepare cost estimates at the planning stage of a project, when definitive design information to support detailed analysis is not available. Credible planning-stage cost estimating requires the development of at least a conceptual design to support reasonable comparisons with past projects.

Complete plant automation systems are multi-faceted, complex and varied. As noted in a previous chapter, rule of thumb costs are not reliable. While a gross comparison of similar automation projects may be generally informative, details vary enough from project to project to make such comparisons insufficient for most cost/benefit evaluations. More accurate comparisons can be made by breaking down a proposed project into components. It is easier to find reasonably close comparisons with previous projects at the component level than it is at the overall plant or process level.

The inherent modularity and maturity of automation technology makes it possible to develop preliminary designs based on well defined and quantifiable generic components and cost elements. However, in the planning stage, the degree to which a project can be usefully broken down is limited by the level of design detail that is known at this stage. This limits the usefulness of planning-stage cost estimates, which necessarily require assumptions, simplifications, extrapolations, and often some guesswork.

The good news is that automation technology and design concepts are well established and sufficiently understood to support reasonable planning-stage cost estimates. This report endeavors to provide a generic automation cost model and cost database that can be used in the project planning stage to make reasonable cost estimates for a wide range of potential water treatment plant requirements.

WATER TREATMENT PLANT AUTOMATION SYSTEMS

Although it would be desirable to arrive at a "typical" or "average" cost of plant automation systems, as described previously, this is difficult because of the wide variety of approaches and components used in a plant automation system. There are a wide range of factors such as plant processes, physical layout, plant location, and access to vendors and services that make it rare to find a plant control system that is identical to another. Usually, automation systems in medium and large size facilities are custom designed and implemented to meet the unique process and utility requirements. Thus, a "typical" control system, with associated costs, does not exist. However, modern WTP control systems also share many similarities and common features.

For the purposes of the construction cost estimating guidelines, Figure 3.1 depicts typical elements of a WTP control system, organized in the following categories based on the functional level of the element and its location in the plant:

- Process Monitoring and Control
- Process Automation
- Plant-wide SCADA

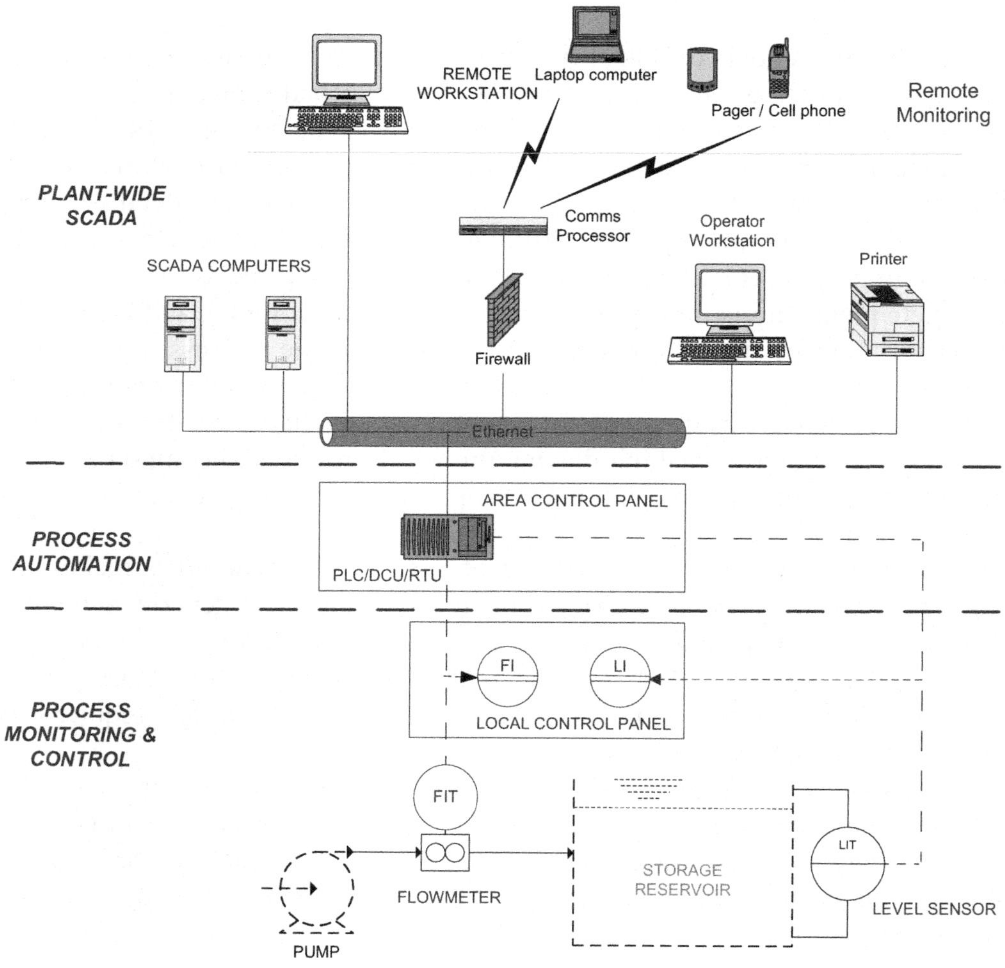

Figure 3.1 Typical WTP automation system elements

Process Monitoring and Control

This category consists of primary instrumentation, including pressure, level, flow, and analytical instruments; primary control devices such as valve actuators, and solenoids; and electrical equipment, including motor control centers, variable frequency drives, and packaged control panels. The types and quantities of devices needed at this level are dependent on the plant processes and capacity, and to some extent, the physical layout of the plant.

Process Automation

The process automation category includes area control panels, local indicators, programmable logic controllers (PLCs), remote terminal units (RTUs), distributed control units (DCU), dedicated operator interface devices; panel mounted recorders, indicators, and single loop controllers. Many of these devices are considered instruments but the intent here is to show generally where they would be located in the plant. Like the process monitoring and control category, the types and quantities of these devices depend on the plant processes, the level of unattended operation, and plant capacity and physical layout.

Plant-wide SCADA

This category includes operator interface workstations, computer networks, printers, SCADA software applications, reporting applications, alarming systems, firewalls, communications processors, and network storage devices. Many of these devices and systems are similar for any water treatment plant, and the equipment costs are only moderately affected by the type of processes and physical size of the plant. Software development costs, including graphic displays, PLC programming, and report development are directly related to the processes used, field equipment count, number of SCADA computers, and plant capacity.

The plant-wide SCADA may also include integration with other utility applications such as data warehouses, laboratory information systems (LIMS), geographical information systems (GIS), electronic O&M manuals, plant optimization, or energy management software. The extent to which these are included in the costs to operate unattended depends on the specific functionality provided by these applications.

Remote Monitoring

This includes devices or systems for transmitting or communicating information off-site from the water treatment plant, such as remote operator workstations; cellular phones; dial-up systems or leased phone lines; licensed and un-licensed radios; fiber optic networks; alarm dialers or wireless alarming systems; closed circuit televisions; intrusion detection and security systems. The costs of equipment at this level are less related to plant capacity and characteristics; however, there is some variation depending on the communications method employed.

One challenge to quantifying the equipment costs is that plant automation systems are made up of a wide range of hardware and software systems, usually supplied by a number of vendors and manufacturers, and that these elements need to be designed to work together. These unique requirements must be clearly understood when estimating the costs of components and the associated services, and ultimately the total system costs.

COST AND BENEFIT CATEGORIES

It is useful to consider automation costs and benefits in two main categories, tangible (economic) and intangible (non-economic or strategic).

Tangible Costs

Tangible costs include those costs that can be assigned an economic value and that are readily quantified. For automation projects, some of the primary costs include planning, design engineering, procurement, and implementation. Often overlooked, but potentially significant, are post implementation costs including computer software and hardware maintenance and upgrading, staff training, and instrument calibration.

Intangible Costs

Intangible costs include technology or operational risk; changes to operating procedures; employee concerns; and change management costs as an organization adapts to new technologies and practices. Often such costs are difficult to quantify, but they do represent a potential impact to the organization.

Tangible Benefits

Tangible benefits can include reduction in labor costs as a result of the facilities being operated automatically or unattended, reduction in travel time to remote facilities; reduction in chemical costs as a result of better control; operational improvements resulting from automation; ability to install additional processes or support plant expansion using savings from avoided costs; labor reduced energy costs through automated load shedding or shifting strategies such as off-peak production and pumping schedules. Other tangible benefits of automation include reduction in labor for data collection and report development, and better data to support equipment maintenance.

Intangible Benefits

Intangible benefits are items whose economic value is difficult to determine. Such benefits such as more consistent quality of treated effluent; streamlined regulatory reporting; better data collection for engineering, planning, and documenting performance: enhanced monitoring and security; rapid response to process upsets or alarm conditions; reduced technology risk and improved operator morale.

CONTROL SYSTEM PROJECT PHASES

It is useful to consider the specific cost and benefit categories in terms of when they might be expected to occur in the life-cycle of a project. Figure 3.2 provides an overview of the stages of a typical automation project. The majority of costs occur in the Planning, Design, and Implementation phases, with some additional costs occurring in the post-acceptance period. The majority of benefits occur after the systems have been commissioned and satisfactorily started up.

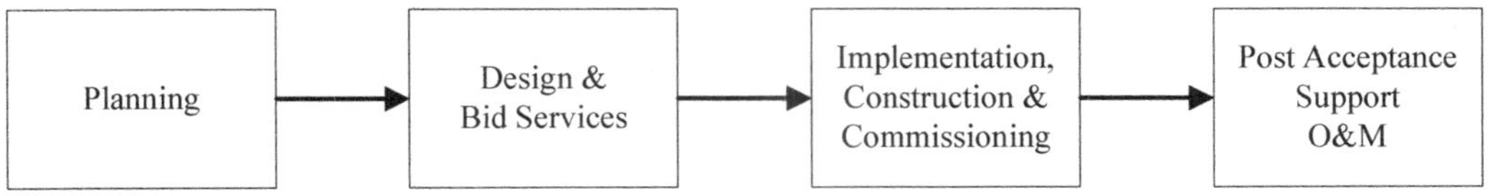

Figure 3.2 Stages of a typical automation project

Procurement Approaches

Several different approaches can be taken to procuring automation projects including:

- Design-Build, whereby the project requirements are defined in general terms and a single entity is employed for the detailed design, procurement, and construction;
- Design-Bid-Build, whereby the details are defined, the plans and specifications are used to procure the services of a prime contractor, and all procurement and installation is done under one contract between the owner and the contractor; and
- CM-at-risk, whereby the construction manager, engineer, owner, and contractor share in the risk of a project. Variations can include Design-Bid-Build, with a third party providing the software integration services under a professional services agreement.

Each approach has a unique set of benefits and drawbacks, depending on the type of project. The cost estimating methods presented in the report are focused on the Design-Bid-Build approach, which is common for water utility projects.

AUTOMATION COST ESTIMATING

Planning

The costs associated with planning can vary depending on the complexity of the project and usually fall into the following categories:

Utility Staff Costs

This cost is incurred by the utility's or owner's engineering, IT, operations and maintenance staff in documenting existing systems and defining the needs for improvements. The process of gathering data on existing facilities, existing control devices and system documentation often involves significant effort.

Engineering

This cost typically includes preliminary planning and engineering analysis to define the overall scope and objectives of the project. It may include the costs associated with engaging an engineering consultant for facilitation or development of a SCADA master plan, part of an integrated technology plan, or a detailed implementation plan. The costs associated with planning can usually be estimated on a time and materials basis.

Design

This item may include engineering costs associated with the preparation of the plans and specifications for competitive procurement. Design costs associated with preparation of plans and specifications may be lower where competitive procurement is not required. For design-build type projects, this item can also include detailed design and engineering work, some of which may be performed by the design-build contractor.

The design engineering fees for automation projects can be estimated in a number of ways. These include the level of effort and expenses associated with preparation of plans and specifications; or based on the number of drawings expected to be produced, with a corresponding average cost per sheet; or a percentage of the total project construction cost. It is worthwhile to note that some of the industry benchmark engineering cost data used for multi-discipline projects may not be applicable to projects that consist primarily of automation improvements. Automation projects tend to have smaller hard costs for construction and equipment, but because of their complexity require a significant amount of engineering effort.

Time and Materials

This approach to estimating design fees is based on developing a detailed breakdown of scope and level of effort for design activities and including the associated costs for direct expenses.

Drawing Count

Another common approach to estimating engineering fees is to base it off the quantity and complexity of drawings required to define the work. This approach can be used to adjust the fee estimate to account for client specific documentation and degree of detail. Many agencies have established drawing and design standards that require a significant amount of detail and effort. Typical engineering drawings required for automation projects include the following:

- Network Block Diagrams
- Process & Instrumentation Diagrams (P&IDs)
- PLC Layout Drawings
- Control Panel Details
- Interconnect Diagrams
- Site and Floor Plans
- Electrical Plan Drawings
- Cable and Raceway Plans and Schedules

- Elementary Drawings
- One-line Diagrams
- Installation Details

The level of effort for these drawings can range widely. Budgetary level of effort numbers can range 12 to 48 hours per sheet, including engineering and CAD depending on the complexity level of the drawings.

Percent of Construction Cost

The American Society of Civil Engineers (ASCE) has published fee curves that relate engineering fees to the cost of construction, for a number of areas of construction (*ASCE Manual of Practice Number 45*, 2003). The guidance provided by the ASCE indicates that these types of fees should not be interpreted as absolutes but can serve as a starting point for negotiation of fees.

Carr and Beyor, 2005, analyzed the drawbacks of these fee tables and the problems associated with "a percentage of construction" approach. They point out that many government agencies have adopted this type of approach, but data indicates an erosion of engineering fees over the years, since the fee tables have not been adjusted in a consistent way or adjusted for inflation.

PSMJ 1998, survey results reported the lowest use of the percentage of construction cost method in the areas of water, wastewater, sewers, roads and bridges (linear construction). Respondents to this survey reported using percentage of construction 60 percent of the time for fee computation, with approximately 45 percent of the lump sum or percentage contracts executed. This information was for multi-discipline projects and not specifically automation projects.

The relevance of this information to estimating control system engineering costs is that WTP automation projects may be part of a utility's overall capital improvement plan or part of an overall plant expansion, retrofit, or rehabilitation project. The same benchmarks or method of fee estimating that is used for the large capital projects at the time sometimes may be used to estimate engineering fee for automation.

Carr and Beyor 2005, present recommended curves for public works projects in several different categories. The data indicates that the more complex the project the greater the fee percentage. Some of the information they presented is applicable to complex building projects however that data was not necessarily applicable to automation projects.

The limited available data indicates that these percentages do not apply to stand-alone automation projects. Anecdotal evidence supports the idea that as a percentage of construction costs, the engineering fee tends to be higher than indicated by the standard ASCE tables. The design of automation projects is typically complex and involves detailed descriptions of project requirements, more drawings and specifications and construction costs are lower than those for civil or building projects that include a large expenditure for materials. It is not uncommon for automation project design fees to exceed 30 percent of the total project cost.

Bid Services

This item consists of administrative and engineering costs associated with advertising, pre-bid conferences, bid evaluations, negotiations and contract award. The costs of these activities can be estimated on a time and materials basis.

Construction Phase Support

The following are the typical engineering services utilized during the construction phase of a design-bid-build project.

Implementation and Construction Services

Construction services consist of efforts during implementation and construction to administer the construction contract and to verify and validate that the project is delivered in accordance with the plans and specifications. This can include submittal review, response to requests for information, change order preparation, review of progress, commissioning and test witnessing.

Software Engineering and Integration

The costs for software and integration services are part of the cost estimating model described above. However, a number of factors including the difficulty in procuring software integration services under a low bid contracting vehicle, have contributed to the recent trend for utilities to find ways to procure these services using a qualifications-based approach.

Many agencies have procured these services as a part of the engineering services or have employed the services of a third party through qualifications-based procurement separate from the construction contract. Having the programming services procured separately from the construction services presents a potential challenge but the importance of performing this work correctly cannot be overstated. These alternative approaches to system programming and integration have been used because of the difficult nature in defining requirements of the work with sufficient accuracy and clarity so that bids can be compared objectively.

Contracting Method Best Practices

Freeman and Prutz, 2004, identified more than 10 best practices for the reduction of the total life-cycle cost of SCADA systems. One highly rated best practice was to engage a professional engineering firm to design the systems under a cost plus fixed fee professional services contract. They noted that contingencies are inevitable in a major SCADA design and this type of professional services contract provides a mechanism for adjusting the design budget. However, the overall contract often includes a fixed fee or profit. This approach provides an incentive for the engineering firm to complete the project within the contracted budget and time frame, yet provides the flexibility for contingencies and changes.

IMPLEMENTATION COSTS

The most significant cost category for an automation project is the Implementation, Construction and Commissioning phase. These costs are also the most important and are usually the determining factor in terms of whether or not to go ahead with the expected project, even if benefits are to be significant. Since cost estimates are predictions of future costs, the true cost of an automation project cannot be known until the project has been implemented. Thus, cost estimates are informed judgments based on comparing a proposed project with past experience.

The term "implementation cost" encompasses all aspects of work incidental to implementation (excluding design engineering that precedes implementation), equipment, panel assembly and wiring, application development, infrastructure construction, installation, testing and startup, training, documentation, and startup/cut-over.

The central challenge in developing a cost estimating framework for automation projects is to balance detail and accuracy against simplicity and ease of use. Greater detail allows more direct and accurate comparisons between a proposed project and model component costs. Yet, too much detail can be an obstacle to effective use of the model and may not result in a significantly more accurate estimate. Some details matter more than others. For instance, the difference in cost between a valve operator with only limit switches and one with a position sensor may not be significant in the big picture; however, the difference in cost between a 6 inch and a 24 inch magnetic flow meter would be significant.

The following sections describe methods and provide data for development of realistic implementation cost estimates at the planning-stage. The intent is not to provide detailed instruction on preparing estimates for bidding, but to provide guidance in developing a budgetary range of values to support planning. Supplemental information is provided in Appendix C - Cost Database and in spreadsheets included on the attached CD.

Generic Implementation Cost Model

The solution presented in this report is a multi-step model as represented in Figure 3.3 Generic Implementation Cost Model. The model is an Excel® workbook based tool consisting of several separate worksheets. The Component Cost Database spreadsheets contain the cost information including labor rates, expense estimates, equipment unit pricing, and software unit pricing, that is used in the higher level model spreadsheets. These costs are based on 2006 prices. The level of detail in the database is intended to be sufficient to account for significant cost differences, but is not so detailed as to require fine-grained analysis and design decisions.

Automation Package Spreadsheets are organized by unit process area and system wide control system elements, and cross reference the unit pricing information in the Component Cost Database. They provide the framework for developing cost estimates for automation systems that are typically found in water treatment plants. To develop the estimate, the user has to add project-specific information.

The Project Summary Spreadsheet combines the unit pricing information and the project specific-requirements to develop a roll-up and summary of the estimated costs. The costs are provided as a range of expected values.

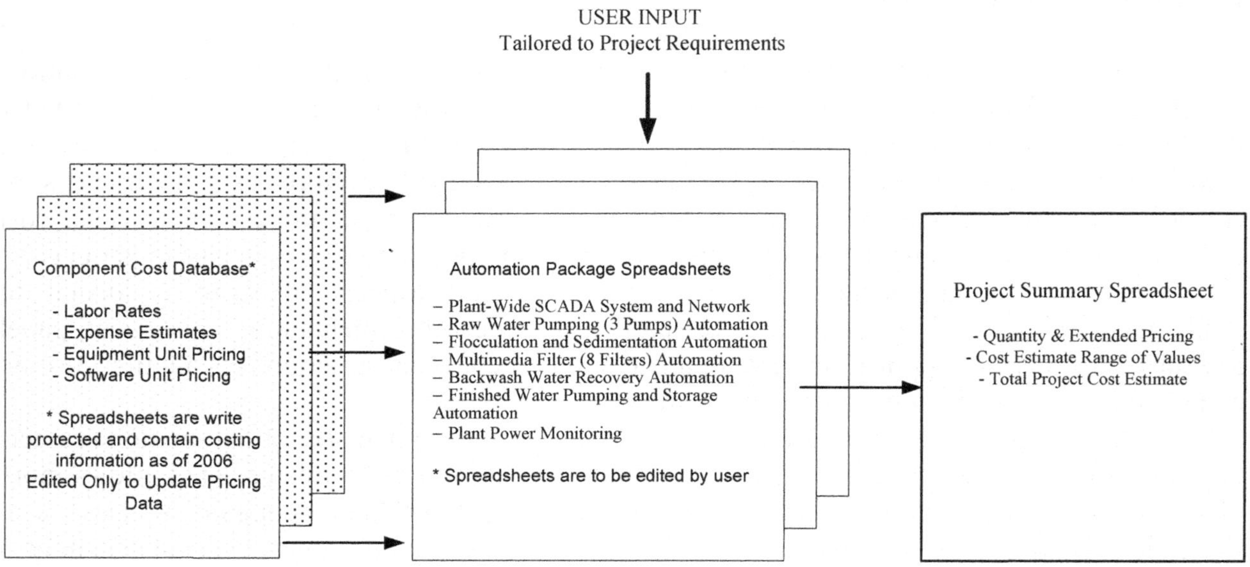

Figure 3.3 Generic implementation cost model

Automation Package Spreadsheets

The generic automation packages in the cost model are representative of typical automation project elements needed to achieve complete automation for a conventional treatment plant. The cost model in Appendix C includes the following generic automation packages:

- Plant-Wide SCADA System and Network
- Raw Water Pumping Automation (Up to 3 Pumps)
- Flocculation and Sedimentation Automation (Up to 2 Process Trains)
- Multimedia Filter Automation (Up to 8 Filters)
- Backwash Water Recovery Automation
- Finished Water Pumping (Up to 3 Pumps) and Storage Automation
- Plant Power Monitoring
- Security Systems

The Automation Package Estimates are comprehensive, that is, an attempt has been made to include all significant component costs. Diagrams in Appendix C illustrate the model configuration, and a spreadsheet lists the components, quantities, and a range of unit and extended costs. These are nominal model estimates, which can be adapted and extended to address the particulars of specific projects.

The cost data spreadsheets on the CD were originally password protected and care should be taken in editing the spreadsheet source data in the event the spreadsheets are unprotected. The password to unprotect the sheets is case sensitive and originally was set as "AWWARF."

Associated components have been grouped into functional sets where the quantity and type of individual components is not likely to vary. For example, the PLC control panel for a process area will have a base set of components and implementation costs that will be the same for all applications. These are summarized as a single component set (line item). However, the quantity of input/output points will vary, so these are listed separately to enable users to account for quantity differences.

Component Cost Estimate Database

The component costs in the Generic Automation Package Estimates are referenced from the values in the Component Cost Estimate Database spreadsheets, also in Appendix C. These database spreadsheets provide cost range estimates for a variety of generic automation components, based on 2006 prices. Components are grouped into generic categories. Subcomponent costs are also shown, which may be adjusted to fit individual different cases. The model for this spreadsheet database is illustrated on Figure 3.4.

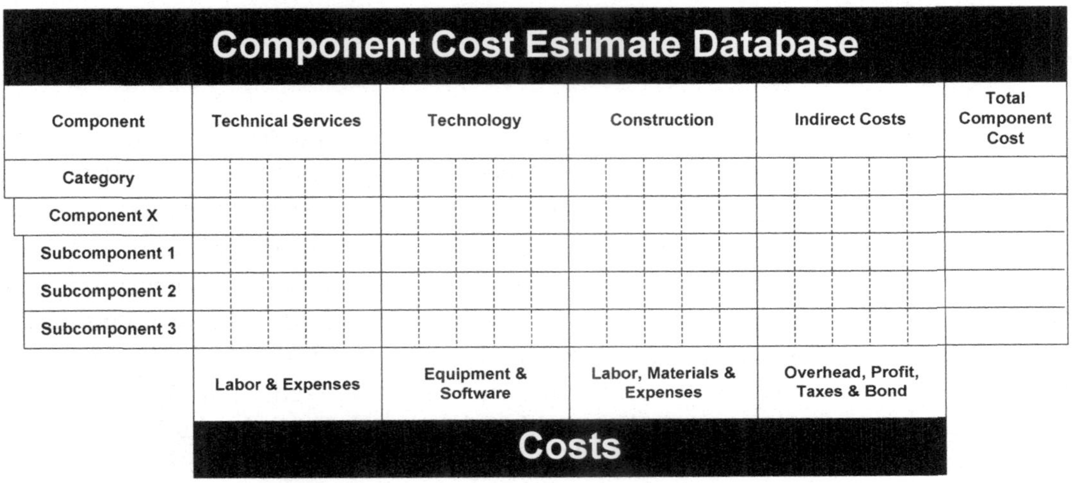

Figure 3.4 Component cost estimate database model organization

Each component cost is derived from estimates of the contributing cost elements, which include the following:

Direct Costs

These costs apply directly to the project work, and will not be incurred if the work is not performed.

Technical Services

Technical Services include the labor and expenses associated with project management/administration, submittal preparation, subcontractor management, detailed implementation design, software development, systems configuration, software development, application programming, equipment rack/panel assembly, factory and field testing, technical supervision of installation, field startup and cutover, documentation development, and staff training. Craft labor associated with these activities is included under Construction.

Technical Services can be provided by a control system integration contractor, an engineering consultant or the owner agency's technical personnel. The cost of these technical services is in addition to the engineering costs associated with the initial planning, design, construction phase services, and construction management and administration.

Technology

Technology cost is the purchase cost of automation equipment and software licenses. The costs in the database for each component or subcomponent include all ancillary parts needed to integrate, install, and operate the component. Electrical and mechanical equipment such as VFDs, MCCs, valves and actuators are not specifically included in the estimating spreadsheets but would need to be accounted for separately, if needed for the project. Sales taxes are not included here, but would be included under indirect costs.

Construction Cost

Construction cost includes labor, materials, and expenses associated with work typically performed by craft labor. For automation projects, this includes mounting and installation of equipment, sampling lines, or process connection piping/tubing for instruments; and installation of conduit, wires, and cables. No allowance has been included for demolition. Construction or modification of process piping or structures has not been included in the estimates, nor has installation of final process control elements, such as valves or gates. The installation cost of in-line devices and process taps includes the associated piping modifications. The installation cost for final control element actuators or controllers, such as motorized valve operators or variable frequency drives, have not been included. Material costs cover the bulk raw construction materials (pipe, tubing, conduit, wire, cable, mounting channel, hangers, etc.).

Indirect Costs

Overhead and Profit

Overhead includes the general and administrative costs that do not apply directly to performance of the project work, but which are incurred by the implementing organization just to operate, regardless, whether or not any project work is performed. Profit is the reward a contractor or consultant receives for the business risk associated with performing the project work.

Taxes

In some jurisdictions, certain aspects of the project may be subject to sales or use taxes. A place is included in the model to apply these, but no amounts are included in the estimate database.

Payment and Performance Bonds

This category covers the fee paid to a surety in exchange for underwriting Payment and Performance Bonds. Bond costs apply only when the owner agency requires the implementing contractor to post a bond. Because this is generally the case for construction contracts the estimate database includes representative costs for bonds.

The labor rates in the component cost estimate database include direct wages/salaries plus the cost of employee benefits and payroll taxes. This is commonly referred to as the "burdened labor rate." The basic unit of labor used in the database is work hours at normal rates, that is, without overtime charges. Low and high labor rates are provided. This range represents the regional variations in wages throughout the United States, and the range of skill levels that may employed. A labor rate table is in Appendix C. The rates should be adjusted as appropriate when applying this model to particular cases, to reflect local labor cost and escalation subsequent to mid-2006, when this model was created.

Allowances are made for expenses that result directly from performing the project work, and include the following typical expenses:

- Temporary Jobsite Office Rental
- Jobsite Office Utilities and Services (power, telephone, housekeeping)
- Office Supplies and Equipment (Paper, copier, FAX machine, computer)
- Postage and Express Delivery
- Vehicle Expenses (rental, gas, servicing)
- Travel (airfare, lodging, meals, incidentals)
- Tool Rental and Special Testing Equipment (large/expensive tools only; typically does not include small hand tools)

IMPLEMENTATION COST ESTIMATING

This section provides guidelines for developing a planning level estimate of automation project implementation costs using the data in Appendix C and the spreadsheets on the CD. The recommended steps include:

1. Define project objectives, scope, and high-level requirements.
2. Develop a conceptual level design.
3. Divide the project design into work packages or unit process areas with components that correlate as closely as possible to the Automation Packages presented in Appendix C.
4. Organize the work packages into estimate worksheets similar to the Automation Package Estimate tables.
5. Adjust quantities of each component to reflect the requirements of the project.
6. Identify components of work packages that are reasonably similar to those presented in the Component Cost Estimate Database.
7. Apply Component Cost Estimate Database total costs to each work package component, making all appropriate adjustments in Component Cost Estimate Database as described above under Factors Affecting Cost.

8. For components not represented in the Component Cost Estimate Database, locate other sources of cost data, such as contractors, consultants, and vendors. The best source for equipment purchases is a manufacturer or its local representative. Construction cost estimating guides, such as R.S. Means or Richardson, are usually sufficient for planning level estimates.

The final step is to derive the total implementation cost of the project and to validate it. It may be appropriate to have the estimate reviewed for errors by another person. How does the project total compare with the total costs of similar projects? If it does not seem to be consistent with the reader's or reviewer's experience or judgment, look for mistakes or erroneous assumptions.

The cost models are based on available nominal cost data for mid-year 2006. The labor rates, material prices, and expenses will have to be adjusted for inflation over time, and the labor hour estimates may have to be adjusted for factors affecting productivity or the complexity of the work.

The Component Cost Estimate Database provides a range of costs for each cost element. For labor estimates, the low and high work hours reflect the nominal range of complexity or difficulty. For equipment and material items, the cost values represent the range of quality, performance or capability of the different products. When using this data it is necessary to decide how the requirements of a particular project relate to the model.

It may be reasonable simply to factor the project total up or down to compensate for apparent bias in the estimate or to account for factors not represented in the estimate detail. It may also be appropriate to estimate separately specific project or client requirements such as programming standards, special testing requirements, additional training, or additional warranties.

The approach is demonstrated with an example implementation cost estimate in Appendix C.

ADDITIONAL FACTORS AFFECTING COST

Extraneous factors, can cause project costs to be higher or lower than estimated. These factors must be considered when applying the model to specific cases. Some factors require judgment to be applied such as assessing the relative complexity of controls or site specific conditions that will increase the difficulty of construction work. Others such as labor rates and material/equipment price inflation involve objective adjustment for economic changes. Factors affecting costs include the following:

Market Conditions

- Who will perform the work (contactor, consultant, in-house)?
- Local labor market and wage rates
- General economy (are contractors busy or not?)

Working Conditions

- State of Existing Automation
  - Reuse without change

  - Extension/expansion
  - In-place replacement/upgrading/enhancement
  - Demolition and rehabilitation
- Known or Potential Interferences
  - Ongoing operations
  - Predecessor projects
  - Concurrent projects
- Operational Constraints
  - Seasonal weather changes
  - Limited outage windows
  - Live cutover
  - Hazardous or corrosive locations or processes

Automation Requirements

- Levels of Control
  - Field manual
  - Local automatic
  - Local auto-manual
  - Remote automatic
  - Supervisory monitoring and control
  - Advanced control

Procurement Methods

- Automation as Part of General Construction Contract
  - Design, low bid, build
  - Design, pre-qualify integrator, low bid
  - Design build

- Automation as Prime Contract
  - Design, bid, build (low bid award)
  - Design, prequalify integrator, low bid
  - Design, prequalify integrator, evaluated proposal
  - Design build, evaluated proposal

Reliability and Expected Life

Life-cycle cost is the total of all capital, operating, maintenance, training, and replacement costs amortized over the life of the system. An important part of evaluating the total cost of automation, particularly after implementation, is the expected life of the components. Some components, such as computers and software have a considerably shorter life expectancy than, for example, valve actuators and wiring. These differences should be taken into account when preparing a comprehensive cost analysis.

It is important to understand the relative life expectancy of components when estimating costs for the life of the project. Table 3.1 provides some information on the life expectancy of typical control system elements.

An important consideration in an economic cost-benefit analysis is the project life-cycle or span. In this report, a life-cycle of 10 years was chosen as representative for automation projects. Depending on the specific elements of the project this value can be modified.

POST-ACCEPTANCE COSTS

The post-acceptance period is defined as the time after final acceptance, or close-out, of the project, which may also be the start of the warranty period.

Maintenance Costs

To realize its purpose, plant automation must be used and must be maintained in good working order. Effective automation reduces operating costs by eliminating the need for human involvement in process operations, and by improving the efficiency of the processes through regulating chemicals and electric power. Such cost reductions are balanced by the maintenance costs of the automation systems and equipment.

Maintenance of automation systems involves both routine servicing and corrective maintenance or repairs. Routine maintenance includes activities such as cleaning, calibration, setting up new system users and making archive data images. The costs of routine maintenance can be defined and are predictable, and can therefore be reasonable estimated. Corrective maintenance is needed as a result of wear, aging, degradation, or random failure and is more difficult to predict with accuracy. However, over time the probability that corrective servicing that will be needed can be estimated based on the service history of the products in use, or can be derived from the historical performance.

Table 3.1
Life expectancy of typical control system elements

Equipment	Typical Economic Life (years)
PLC and DCS Hardware	15 – 20
Instruments	15
Computers (for HMI)	3 – 6
Operating System Software	3 – 6
VFD	15 – 20
Motor starters	25
Communications hardware	10 -15
SCADA software (with periodic updates)	6 – 10
Valves and actuators	25
Pumps and motors	15 – 20

Modern automation technology is very reliable and requires little preventive maintenance. Plant-wide SCADA systems do require a moderate amount of routine maintenance to ensure that data are preserved, user accounts are created as needed, and to monitor the overall health of the system.

The complexity of the SCADA software and configuration makes these systems susceptible to damage from faulty hardware processing or human error. Both can have subtle effects that may not cause immediate problems, but will in time result in abnormal operation.

Proper routine maintenance can minimize the risk and effects of such damage, and can identify and correct its cause before it affects plant operations. Because they are exposed to wet processes, instrumentation sensors require regular servicing and are more susceptible to degradation and failures than dry electronic components.

An effective automation maintenance program must include technicians and service engineers with specialized skills to perform routine and corrective maintenance on the equipment installed, and a stock of supplies and replacement parts readily available.

Routine maintenance costs that should be considered in estimating ongoing maintenance costs include the following:

- System Administration and Maintenance
  - User account setup/maintenance
  - System image and data archive backup
  - System status monitoring and health checks
  - Hardware replacement and upgrading
  - Software licenses
  - Software upgrades/patches
- Telecommunications costs including cellular service, T1 service costs, Internet access, and leased lines
- Component inspection, cleaning, filter replacement
- Instrument and controller health checks and calibration
- Routine analyzer probe replacement
- Consumable supplies
  - Printer paper and ink cartridges
  - Data archive media
  - Filters
  - Replaceable Probes

Corrective maintenance costs are a function of the inherent reliability of each component and the service conditions. Component manufacturers can usually provide Mean-Time-Between-Failure (MTBF) data for their products. MTBF is the average time a product will operate without failure. This data may be derived by calculation, or may be determined by the service history recorded by the manufacturer. MTBF can be used to calculate the probability that a component will fail in one year. The average annual corrective maintenance cost is derived by multiplying this probability by the cost to repair it. The cost of repairs can vary from the cost of labor involved in finding a problem in software or configuration and making a small change, to complete replacement of a component.

The Component Cost Estimate Database in Appendix C includes ranges of the cost of both maintenance labor and total maintenance. Routine maintenance is based on typical manufacturer's recommendations for preventive maintenance and system maintenance. Corrective maintenance costs are based on estimated MTBF and the estimated costs of corrective maintenance labor and component repair. The estimated maintenance cost for a given automation project is simply the sum of the estimated maintenance costs of the project components.

Spare Parts Inventory

Typically, automation system components are so critical to unattended operation that it is imperative to complete corrective maintenance as soon as possible when a failure occurs. This requires immediate availability of replacement parts from a readily accessible and well-stocked source. The establishment of a spare parts inventory represents an investment that should be considered as part of evaluating the cost of automation. To assist in determining the cost of spare parts, Spare Parts Inventory $ is included in the Cost Database.

TOTAL PROJECT COST

Determining a budgetary estimate for the complete project cost requires adding the costs of all phases including planning, engineering, implementation, construction phase support and post acceptance support. Chapter 6 provides an approach to considering total project costs.

ESTIMATING THE BENEFITS

Several literature references identify the benefits associated with automation. However, few references provided an approach to quantifying tangible benefits. Many of the benefits were expressed in terms of overall savings as opposed to specific savings on a per unit basis that would make it easy to estimate savings. In general, the tangible benefits associated with automation include:

- Labor savings
- Chemical cost savings
- Energy cost savings
- Maintenance cost savings

In the absence of extensive industry benchmark data, estimating the benefits of automation can require an in-depth analysis and making assumptions. Estimating the benefits associated with automation improvements, in particular those associated with moving from attended to unattended operation is difficult and requires a thorough understanding of the impacts associated with such change. The approach recommended for existing facilities is to base the estimated savings as a percentage difference from an historical baseline. The following are sources of data for estimating benefits:

- Utility Historical Baseline Data - percentage of savings
- Production Data – will there be a percentage change in production?
- Energy Cost Data – utility bills
- Utility Labor Data – payroll and existing staffing levels
- Chemical Costs - invoices

- Maintenance Data – invoices for materials, labor related to automation
- Utility case studies
- Industry Benchmarking data (limited)
- Any planned facility improvements that might add control equipment, rotating equipment, pumps, motors, and any estimated additional staffing level requirements

It is assumed that the analysis is being conducted to assist in determining whether a project or a change in operations should be undertaken, and what the economic and intangible impacts might be to the organization. Chapter 5 provides a review of the typical process areas in a water treatment plant and identifies areas where benefits may be realized.

LIFE-CYCLE COST BEST PRACTICES

Water and wastewater utilities are very interested in designing, implementing, operating, and maintaining equipment to achieve the lowest overall life-cycle cost. Recent industry initiatives embracing asset management-based principles also indicate awareness that life cycle costs and benefits are certainly important but that they need to be balanced with meeting appropriate service levels and understanding the risks associated with failure to achieve the desired levels of service.

Several references provide information on approaches taken by utilities in implementing automation for water treatment plants and identify factors that impact the cost and success of projects.

Freeman and Prutz 2004, identified 25 best practices for reducing the life-cycle cost of SCADA projects for water and wastewater utilities. The following are the top 10 of these practices in the order of ranking:

- Open systems architecture; Modbus support, no proprietary solutions
- Prior successful SCADA projects by the selected engineering design team
- Supplier support history
- Professional preparation of system specifications
- CSIA registered supplier with 5+ years of experience on similar projects
- Flexible ongoing training
- Cost-plus professional services contract
- Supplier's ability to design, assemble, start up, and service as a single source
- Request for proposals process
- Supplier prepaid contract for post-startup user development, training, and maintenance

Several other lower ranked best practices were identified in their report, many of which focused on technical aspects or detailed design issues.

SUMMARY

This chapter considered a number of elements that are important to developing a complete picture of the costs and benefits of WTP automation. The goal of this chapter was to provide a review of the major components of typical water treatment plant automation systems;

to describe the cost and benefit categories; and to present a detailed approach to estimating construction costs; the system operation and maintenance costs; the considered potential benefits; and the strategies for minimizing life cycle costs. Understanding this information will assist in developing the business case analysis in Chapter 6.

CHAPTER 4
AUTOMATION CONSIDERATIONS

The decision to operate a plant in an unattended manner hinges on the ability of automation technology to perform predictably and dependably. Utility managers need to have a high level of confidence that the automation will make predictable control responses, and in the event of plant upset, make appropriate control responses. If abnormal conditions persist, the automation system must properly notify the correct individuals in the event operator response is necessary.

This chapter provides an overview of some of the technology considerations for unattended operations plus an approach to identifying potential risks and mitigation strategies. The chapter closes with a list of minimum automation/unattended operations considerations and recommendations.

WATER TREATMENT PLANT AUTOMATION COMPONENTS

Figure 3.1 provided a simplified schematic of a typical WTP control system. This generalized diagram shows the major classifications of equipment that comprise a plant automation system. The figure is organized with monitoring and field instrumentation devices on the bottom. Field devices include primary instruments, transmitters, and final control elements like valve controllers, pumps or variable frequency drives.

The process automation level in the middle of the figure includes local control panels with single purpose displays or process indicators, alarm indication or dedicated process loop controllers. Each major process area of a WTP typically has an associated local control panel. The local control panel typically contains both hardwired devices for backup control and monitoring and a Programmable Logic Controller (PLC) or Distributed Control Unit (DCU) for automatic control. These controller units serve as the connection point for field device signals. These signals are commonly referred to as Inputs and Outputs (I/O).

The plant wide SCADA system level depicted above the process area level requires a computer network to connect the PLCs/DCUs in the plant process areas to a central SCADA computer. The computer network may be proprietary or a standard computer network such as Ethernet. The central SCADA computer may be configured as a dual redundant system for reliability. Workstations for operators, supervisors, management, and other staff are also connected to the computer network. For larger systems, a data archiving computer called a historian may be included.

The central SCADA computers include alarm notification functionality such as dialers, remotely connected operator workstations, pager or cell phone alarming or a connection for remote access through portable or laptop computers. Connections from remote computers must be routed through a device called a firewall to ensure data security. This remote monitoring level is an integral part of any system used for unattended plant operations.

RISK AND FAILURE ANALYSIS

As described above, automation systems are comprised of numerous interconnected elements that need to function correctly in order for the control system to operate reliably. The risks associated with automation systems not functioning reliably can be considered from three perspectives as follows:

- Hardware
- Software
- Data

The hardware risks range across all of the elements described above from field devices to top end computers. Software risks cover the operating systems, applications, and network/communications elements. Ultimately the system gathers, processes, and acts on data which is core to WTP operations. Chapter 5 describes many of the process specific considerations of WTP operations.

Risk, Reliability and Failures

Hardware

The primary risk associated with hardware is simple failure. Considering the field devices first, the risk is that the field device will not perform in the expected manner. These devices operate in the most corrosive, demanding physical environment there is. They are exposed to water, chemicals, process and physical stress, vibration, electrical noise and power surges. These factors increase the likelihood that the field devices will simply not function, or, even worse from a reliability perspective, that they will provide misinformation to the automation system.

The communications network can be exposed to similar risks, but the potential for failures is less since the network is primarily cable, either copper or fiber optic. The network interface equipment is addressed below. Local control panels, while they are distributed in the process areas, are installed in less hazardous areas such as equipment rooms. The primary risk to the overall panel is loss of electrical power. The equipment housed within the panel is subject to similar risks as the field devices. The top end equipment, primarily computers, is housed in conditioned spaces that are relatively benign.

The risk associated with all computing and hardware devices follows a typical three step failure progression. Ebeling 1997, describes this for equipment in general, in terms of a "bathtub curve." The typical failure stages start with a burn in period, where failure rates tend to decline and early failures occur due to manufacturing defects and related failures. For automation equipment this burn in period typically is in the 3 month to one year time frame.

This is followed by a useful life period where failure rates stabilize but random failures can occur due to environmental or chance events, like system overloads or lightning strikes. This useful life period can be in the 6 month to 7 year time frame or beyond, for automation equipment.

The curve progresses to a final period where probability of failure increases to the point where devices fail due to wearing out. This can include failures due to fatigue, corrosion, cyclical failures and simple aging.

Software

There are three areas of risk associated with software. The first is the operating systems that enable the computing devices to function at any level. Failure of this software is dramatic, the equipment fails to operate. The second software risk is the applications software that runs once the devices are operating. The risks here are more subtle and unpredictable; an application can run properly for months, or even years, until an unexpected combination of events uncovers a flaw in the application system and lockup is the most likely failure mode.

The third area of software risk is the configuration of the control logic, graphic displays, reports, and other software features that are unique to the WTP. Occasionally these failures can result from a set of circumstances that were not encountered or tested for that result in failures. This can result in logic failures and failure modes include failure to operate, and resulting process upsets and potential non-compliance.

Data

Data is the finished product that automation systems create. Risks to be considered related to data include accuracy, timeliness, availability and security. The operational risks include failure to operate, and resulting process upsets as well as the potential for regulatory non-compliance.

Automation System Reliability

Considering the wide range of elements that comprise an automation system and the high levels of interaction between these elements, it is difficult to devise a single indicator of system reliability. System hardware elements such as field devices, computers, and other computing devices have inherent reliability factors that the utility cannot influence. Similarly, the operating systems and core applications are standard products from vendors who have the responsibility to make them reliable. The area where the utility can influence reliability is through the configuration of the system hardware, referred to as system architecture and in the configuration of the logic, graphics, and reports.

A more common performance measure for these systems is availability, the gauge of how much of the time in a year the system is performing its assigned functions. Availability is expressed as a percentage by taking the number of hours the system is operating properly divided by the number of hours available in a year. Ultimately, this measure reflects the reliability of the individual components as well as effectiveness of the configuration into which they are placed. This is also a typical way that performance requirements are specified for systems.

AUTOMATION DESIGN RELIABILITY CONSIDERATIONS

With an understanding of automation reliability, risks and failure modes we can consider mitigation strategies that will increase the reliability of the automation improvements being planned to support unattended operation. The following section focuses on design and implementation considerations for the automation system. Discussions will begin with the most fundamental issue, electrical power and progress through the full range of elements that comprise an automation system.

Electrical Power

Redundancy of power feeds or sources for the WTP is a fundamental design criterion. Without electric power, most plants cannot operate at any level. Without power or an alternative pumps cannot run, valves and gates cannot be operated and the automation system cannot gather data nor enable operators to take control actions. Even with redundant power feeds, the distribution of power throughout the plant should be carefully designed to ensure that individual breaker trips do not de-energize entire process areas. Standby power from engine-generators or other sources must be considered and deployed as appropriate.

Presuming an appropriate level of reliability of the overall power system, the WTP design must include batteries and uninterruptible power supplies (UPS) for critical loads. These loads may include field devices, manual controls, and the automation system equipment. UPS equipment must be sized to enable an orderly shut down of the WTP as a minimum. The redundancy of power supplies within control panels should be evaluated to avoid the case where one external breaker trips causing complete loss of power to the automation systems within the panel.

The UPS equipment should be monitored by the automation system so that any UPS failures or malfunctions are monitored and responded to promptly.

Hardware

Field devices such as instrumentation and final control elements tend to be highly reliable. Although there is frequent discussion regarding the use of redundant field devices to increase reliability, there is no evidence that proves that redundancy of field devices really does increase reliability. To the contrary, improperly implemented redundant field devices can actually reduce reliability. Schemes that use redundant field devices that require manual switchover or complex software logic can be prone to failure. The perceived additional reliability can be reduced by complex redundancy schemes.

Initial design of field device installations is the first opportunity for increasing reliability. Instruments must be properly selected and applied for the service they will provide. They must be installed to minimize exposure to the elements, abuse, and isolation from vibration, process fluctuations, and electrical surges. The best approach to ensuring long term field device reliability is a rigorous maintenance management approach that includes periodic inspection, testing, calibration, and replacement of these critical automation system elements.

Communications Network

The communications media (copper or fiber optic) used for control systems tend to be highly reliable. As with field devices, the initial installation sets the baseline for reliability. The media must be installed in proper raceways and protected from physical damage. Terminations and the network equipment must be installed in appropriately rated enclosures to prevent physical damage, protect from environmental hazards, and protect against unauthorized access. Network management software should be utilized to monitor network performance and highlight degradation of functionality.

Local Control Panels

Similar to field devices, the local control panels must be selected and designed to function in the environment where they will be installed. Enclosures with proper NEMA ratings should be selected and the NEMA ratings should not be compromised by inappropriate cutting and drilling of the enclosure. Within the enclosures attention must be given to proper power distribution and particularly to grounding of equipment and field cables. The UPS equipment is often housed within the enclosure and it must be wired to enable operation of connected equipment when the UPS is in service and also when it fails but AC power is still available.

The controllers, whether PLC, DCU or RTU, are inherently highly reliable devices. The key to high reliability is proper system architecture. Controllers today can perform the majority of the computing processing, so the design should enable "stand alone" operation for each controller. For extremely critical processes consideration of redundant controllers is appropriate. During design, the proper assignment of I/O to controllers must be considered. For example, in a process area with four identical pumps, it may be appropriate to have two controllers, each having the full complement of I/O for two of the pumps, rather than a single controller. This ensures that when a failure occurs part of the process will remain in service.

However, for most processes the best approach to ensure high reliability is to have an appropriate inventory of spare parts and an effective maintenance program to deploy them. For CPU failures, having the proper configuration files readily accessible is a must. If redundant processors are deployed, the failover strategy should be exercised regularly as part of the maintenance process to ensure proper operation when a failure does occur.

Master Control Computers

The master control computer level is where the strategy of redundant equipment can be deployed most successfully. Modern SCADA/Control System applications are designed to operate simultaneously on multiple computers. The cost of this equipment has reached the point where it is more effective to have multiple units in service, properly configured to function as primary and backup to each other. A fully configured spare can also be kept off line for a replacement unit in case of catastrophic failure. The key to minimizing impacts and recovering quickly is backing up the current application files.

SOFTWARE CONSIDERATIONS

The single most important aspect of software reliability is keeping copies of all, current configurations. The hardware will fail, the software will lock up, so the key to success is quick recovery from these malfunctions.

Operating Systems

Windows (Microsoft, WA) is the predominate operating system for master control computers. There are some legacy systems that operate on UNIX or VMS operating systems and there are several systems that are being developed using LINUX. Regardless of the operating system, to ensure high reliability, the utility must update the software periodically and ensure that all service packs and patches are installed. The service packs and patches should be installed and tested using off-line computers to insure compatibility with the SCADA application before full-scale deployment to the online system.

Application Software

The software vendor will issue periodic updates to their core applications occasionally. Each update should be evaluated on an off line system to assess its usefulness, impacts to operations, and stability. Frequently these updates have not been tested with the full range of device drivers, so it is critical to test before deploying.

Configuration Files

As noted above, this software element is the key to highly reliable automation systems. When first compiled, the logic configurations must be rigorously tested. It is important not only to test for proper operation under anticipated operating scenarios (for example, on low tank level the pump starts and the inlet valve opens) but to test for proper software operation when the field devices malfunction (for example, on low tank level the pump starts, but the valve does not "see" the pump start so it does not open causing the potential problems of overpressure at the pump outlet and failure to refill the tank). For facilities that are to be operated unattended, testing is extremely critical since there will be no staff available immediately to take corrective actions.

Similarly, the HMI graphics and any automated reports must be tested for proper operation and approved by operations staff for deployment.

DATA CONSIDERATIONS

Everyone has heard of the expression "garbage in/garbage out" in relation to computer systems. This is the fundamental issue for successful unattended operations. The oft cited garbage is in fact the data and the adage is accurate – wrong data will result in wrong operational decisions. The aspects of data that must be right include:

- Accuracy
- Timeliness and availability
- Security

Accuracy

The first link in the data reliability chain is properly deployed and regularly calibrated field instruments. Proper deployment is a function of design. The devices must be sized properly to handle the full range of the measured variable, the range of the instrument must be set properly to ensure it is measuring the real process situation, and it must be installed to ensure proper sensing of the process variable.

Once all of these issues have been addressed and the device has been commissioned and proven to operate properly, WTP staff must execute the maintenance processes and procedures for this device to ensure continued proper operation. So-called "smart" instruments are able to help in this process by reporting their condition, calibration ranges, and other key parameters over the automation system network. This capability, properly deployed, can increase reliability significantly.

The next data element to consider is calculated data. This important data can be raw data from field devices adjusted by computation (temperature compensated flow) or can be a result of a mathematical computation (totalized flow from multiple pumps).

Timeliness and Availability

For the data to be useful, it must not only be accurate, but it must reflect the time it was collected (time and date stamping) and it must be delivered to the right place at the right time. To attain reliability in this area requires proper network architecture and configuration. All networked computing devices should be synchronized to a standard clock. The U.S. atomic clock standard or timing based on the NIST time standard are commonly used for this purpose. If high resolution time stamping is a requirement, special "sequence of events" equipment may be required. This specialized equipment incorporates high-speed scanning technology that captures time stamps with fractions of a second accuracy.

Once the controllers in the field have captured the data, it must be delivered to the users. The first use of the data is by the operators who are viewing graphic displays. Operators and/or automated control schemes use this data to implement control changes. The configuration of the displays and related database must include a feature to "flag" data that is not timely, that is outside of anticipated ranges, and/or is provided by a device or calculation that is suspect. Less urgent but equally important uses for the data include reporting and trending. Here again, proper time stamping is key to beneficial use of the data for post incident troubleshooting. In the case of regulatory reports, improperly dated data may cause non-compliance.

The final issue for data timeliness relates to operator data entry. One way to help ensure reliability of this data is training for those who do the data entry and business processes and procedures for cross checking and confirming that the data is valid.

Data Security

Data security impacts reliability from two perspectives. First, data security is needed to ensure that accurate data is portrayed and delivered to users and regulatory agencies. The repercussions of someone modifying or destroying data, whether accidentally or willfully, are immense. The validity of data must be maintained as it passes from field devices to controllers to users and ultimately to archival storage. Best practices to ensure data security include encryption, password protection, and user authentication. Also, data file backups must be performed regularly and automatically. Multiple storage locations including at least one off site point must be utilized.

Plant and cyber security is the subject of numerous other water industry studies and as such is beyond the scope of this report.

TREATMENT PLANT RELIABILITY CONSIDERATIONS

Although automation reliability has a key impact on overall WTP reliability there are a number of other considerations that can make the overall facility more resilient to potential failures. Reliability has to be designed into the WTP from the start and an evaluation of the process, mechanical, electrical, and I&C issues can assist in making a WTP more predictable or minimizing problems in the event of automation problems. A brief overview of the areas of plant design that can impact reliability is provided below. A more thorough presentation is in Chapter 22 of the *Handbook of Water Treatment Plant Design*. In that book Mr. Spitko presents concepts related to the following items that will enhance overall reliability of the WTP:

- Multiple sources of raw water
- Treatment process redundancy
- Equipment redundancy within each process
- Multiple power sources; both electrical and mechanical
- Alternate flow paths with processes within and between processes.

The utility specific processes and situations need to be considered in the development of an unattended operational strategy and process considerations.

RISK ANALYSIS AND MITIGATION MEASURES

This section describes many of the automation risks that are associated with unattended plant operations and provides recommendations for identifying potential risks and mitigation strategies. The approach utilized can vary in complexity depending upon the magnitude of the potential risk as well as the magnitude of the consequences. The level of effort required to perform a risk analysis can also vary.

Risk Analysis Approach

One approach to performing a risk analysis can take the form of identifying, quantifying and prioritizing risks in terms of consequences and probability of failures associated with not meeting the desired level of service. The focus here is on automation system elements. The key factors in risk analysis are:

- Probability or Frequency of Failure
- Consequence of Failure
- Level of Risk, i.e. the combination of probability and consequence.

Probability of Failure

Automation system failures can include reliability failures, quality of service failures, maintenance response failures, mortality, or condition-based failures. Good sources of data for identifying the types and probabilities of failure are Computerized Maintenance Management System (CMMS) data, maintenance records, technical literature and other similar information available within the utility. Because of the wide diversity of failure modes, it is useful for analysis purposes to assign a numeric value that characterizes the probability or frequency of failure.

As an example approach, Frequencies of Failure can be assigned a numeric value of 1 to 4 as follows:

1 = Every 5 Years +

2 = Every 1-5 Years
3 = Once per year
4 = Monthly

Consequences of Failure

This involves identification of the potential consequences of failure for each project element. Consequences to consider may include the following direct costs:

- Direct repairs
- Regulatory non-compliance = fines
- Costs due to not meeting level of service
- Increased costs or loss of revenue

To provide a consistent basis for assessing the risks of failure, one approach is to assign each control system element a consequence rating. An example is provided in Table 4.1 where the consequence is assigned a rating from 1 to 4. The potential consequences of failure for each element can be determined by discussion with the utility and identifying the utility specific consequences.

Risk Evaluation

Combining the probability of failure with the consequence of failure of each element and/or equipment can provide a depiction of the significance of the risk. An example automation risk assessment table is provided in Table 4.2. An approach to risk rankings is as follows:

- Extreme – Failure would pose an immediate and extreme risk to providing treated water or result in significant equipment damage. Could result in a boil water notification
- Major – This poses a significant risk or impediment to satisfactory operation of the facilities or systems and may result in facility shutdown
- Moderate – This poses an impediment to satisfactory operation of the facilities or systems
- Minor – This condition poses a nuisance to continued operation of the facilities or systems
- Insignificant – Failure would require no further action

Table 4.1
Consequence table

Rating	Loss of All Water Treatment Ability	Loss of Major Automation System Element	Financial
4	15 Minutes +	1 Week +	$100,000 +
3	<15 Minutes	2 Days	$10,000
2	0	1 Day	$2,500
1	0	4 Hours	<$500

Table 4.2
Example automation failure mode - effect risk assessment

Item #	Element and Failure Mode	Probability of Failure	Effect	Consequence of Failure	Mitigation Measures
1	Total loss of electrical power	3	Flooding, plant shutdown, customer complaints	4	Carbon fueled engine generators and/or pumps
2	Loss of electrical power to top end equipment	3	Loss of visibility of process areas, inability to operate remotely	1	UPS, redundant power sources
3	Loss of electrical power to process area local panel	4	Loss of local control for this process after UPS is spent	1	UPS, redundant power sources, logic configured to fail to last state
4	Loss of electrical power to single field device	4	Loss of visibility or control for this parameter	1	Connect field devices to UPS
5	Hardware – loss of field device	4	Loss of visibility or control for this parameter	1	Configuration flags bad data, configuration stops closed loop control, configuration alerts maintenance staff
6	Hardware – loss of comms Network	2	Loss of visibility or control for isolated elements of automation	2	Redundant, physically separate cable routes, network monitoring flags operators and alerts maintenance staff
7	Hardware – process area panel and controller	3	Loss of ability to control and monitor this process area	1	Hardwired local controls, manual control, backup configuration files
8	Hardware – top end computers	4	Loss of visibility of process areas, inability to operate remotely	1	Redundant servers and workstations, configuration flags failures, backup configuration files
9	Software – Operating System total failure	2	Loss of visibility and control of process areas, inability to operate remotely	1	Redundant PLCs, servers and workstations, configuration flags failures, backup configuration files
10	Software – core application failure	2	Loss of visibility and control of process areas, inability to operate remotely	1	Redundant PLCs, servers and workstations, configuration flags failures, backup configuration files
11	Software – Configuration files	3	Loss of visibility and control of process areas, inability to operate remotely	1	Redundant PLCs, servers and workstations, configuration flags failures, backup configuration files
12	Data – accuracy	4	Inappropriate operating decisions, poor water quality, consumer complaints	3	Configuration flags old data, configuration stops closed loop control, configuration alerts maintenance staff
13	Data – timeliness	4	Inappropriate operating decisions, poor water quality, consumer complaints	3	Configuration flags bad data, configuration stops closed loop control, configuration alerts operations staff
14	Data – security	4	Inappropriate operating decisions, inaccurate reporting, Non Compliance	3	Institute data chain of custody processes and procedures, supervisory oversight of data usage

When considering advanced automation to enable unattended operation, utilities are well served to consider the automation improvements from a risk of system failure perspective. Each utility will need to consider its situation to develop a specific risk profile but it is envisioned that this task will identify major and minor issues that impact the ability to meet the desired level of service of the treatment plant.

Identify and Develop Alternatives

Given a better understanding of the risk profile of the various project elements typically there will be the identification and evaluation of alternatives. These may include redundancy, replace or do nothing. The evaluation of alternatives includes a comparison of life-cycle costs. The asset management based approach then results in an evaluation of a particular project or project element, provides a good profile of the service level expectations, the associated risks and costs.

The reliability considerations presented previously reflect an equipment or software failure perspective. It is important to note that there are also project execution risks. If a project is not implemented properly then the dependability of the facility can be compromised.

BARRIERS TO UNATTENDED OPERATIONS

This section identifies many of the barriers that may be encountered in moving the operation of a water treatment plant to an unattended mode as well as some of the potential mitigation factors that can assist in overcoming the barriers. Table 4.3 provides a summary of the potential barriers and mitigations measures.

Table 4. 3
Barriers and mitigation measures

Barrier	Mitigation Measures
Technophobia	Education
Regulations	Clear Understanding of Regulations
Security	Monitoring, Effective Security Policies
Lack of Staff Training	Training and Education
User Buy-in	Training
Lack of Contingency Planning	Develop standard operating procedures or emergency response plan
Poor Vendor Product Support	Correct Vendor selection
Management Trust	Improve Communications, Enhance Performance
Cost	Determine and articulate the benefits, alternative procurement methodologies, Design-build
Poor Maintenance	Preventive maintenance program, Outsource calibration

(Continued)

Table 4. 3 (continued)

Barrier	Mitigation Measures
Failed Automation Implementation	Construction Contract Coordination, standards development, alternative procurement practices
Equipment and software reliability	Replacement Parts, Configuration and Data file backups, Improved Maintenance
Process or Operational Complexity	Redundancy
Raw Water Variability	Optimization models

RECOMMENDATION SUMMARY

It is recommended that the following WTP automation features be considered:

1. Careful design to ensure no single points of failure
2. Rigorous hardware and software analysis and testing before deployment
3. Comprehensive diagnostics and notifications to operations and maintenance staff including:

    - Process trends
    - Process alarms
    - Equipment warning and failure alarms
    - Instrument failure alarms
    - Control system and communications alarms
    - Security alarms
    - Data quality alarms

4. The ability to notify off-site operators of abnormal conditions if the automation system fails (phone, pager)
5. The ability for off-site operators to remotely monitor the plant status using a graphical interface device or laptop
6. Redundant systems (full featured SCADA backed up by an alarm dialer)
7. Daily backup of data files and weekly backup of configuration files
8. Control strategies designed to fail-safe
9. Control strategies that can switch automatically to backup systems when the primary system fails
10. Control strategies that can automatically maintain water quality under a wide range of flows, raw water and chemical conditions
11. Event reporting
12. Historical data collection and reporting
13. Thorough system documentation
14. In-depth equipment and system operations and maintenance training
15. Security including:

- High level of protection from un-authorized computer access.
- Proper facility security (gates, fences, doors, intrusion detection for sites, buildings, rooms and equipment enclosures).
- Established and enforced security procedures.

CHAPTER 5
UNATTENDED WTP PROCESS SPECIFIC CONSIDERATIONS

This chapter considers the costs, benefits, and risks associated with automation of specific water treatment plant processes, with a focus on unattended operation. Although many considerations for operating a plant unattended are at plant-wide level, it is recommended that each process and related component be individually considered before operating it unattended. This is particularly important when considering the potential consequences of equipment and instrument failures and the mitigation measures required.

While the discussion in this chapter covers only representative processes and the considerations associated with their unattended operation, this can serve as the framework for utilities to consider automation of other processes at their treatment facilities. An additional goal of this chapter is to illustrate the approach that can be used for estimating the costs and benefits associated with automation of individual unit processes.

Although this section does consider plant energy consumption, it is not intended to be a comprehensive evaluation of energy use or its optimization. Nor is it intended to imply that all of the potential savings described will be the result of operating a plant in an unattended manner. However, advanced automation needed for unattended operation can also be used to implement energy saving strategies, such as off-peak production and pumping.

GENERAL CONSIDERATIONS

Plant Operation and Maintenance Costs

Global Energy Partners, LLC 2004, presented operation and maintenance costs for a typical water treatment plant. These costs percentages are in Table 5.1.

As indicated, the three largest cost components are staffing (labor), energy, and chemicals. This Chapter identifies potential benefits in these areas by using enhanced automation.

Labor

Automation can reduce the use of plant operating labor in a number of areas especially the performance of routine repetitive tasks, such as starting and stopping equipment; monitoring

Table 5.1
O&M costs in a typical WTP

Item	Percent of Total O&M Cost
Staffing	35 %
Energy	34 %
Chemicals	16 %
Other	13 %
Maintenance Materials	2 %

Source: Adapted From Global Energy Partners, LLC 2004.

and control of operation of both in-plant and remote facilities; supervision and travel time; on-going process adjustments and calculations for process optimization strategies. If the level of automation provided enables partially or fully unattended operation, the cost savings can be significant.

An estimate of staffing requirements for 5 and 50 mgd water treatment plants is in Table 5.2. This data, obtained from *Water Treatment Plant Design, Fourth Edition*, indicates potential reductions in labor (full-time equivalents) for a 50 mgd plant; and a 5 mgd plant when moving from semiautomatic to fully automated mode of operation. The data did not indicate if fully automatic meant that the plants were capable of unattended operation. As seen in the table, the anticipated reduction in operator full time equivalents is over 50 percent in each case.

The authors indicated that although in general, the more highly automated plants required fewer operators, the higher level of automation required additional instrument technicians. They indicated that the greater the number of treatment processes, the more personnel are needed for operation and maintenance. And, the more spread-out a plant and its distribution system, the more personnel are required to staff it.

Examples of staff reductions achievable through automation are in the case studies in Appendix B. The following sections on process-specific considerations describe some of the impacts that automation can have on labor requirements and the automation improvements that can be made to reduce the need for operator action.

Energy

Although this section reviews WTP energy use, it is not the intent to imply that all of the potential savings described in this section will be the result of operating a plant in an unattended manner. However, advanced automation needed for unattended operation may have a secondary benefit if it can enable energy cost saving strategies such as peak shaving and off-peak production and pumping.

The AwwaRF report *Best Practices for Energy Management* indicates that according to a survey of 19 water utilities, energy costs as a percentage of overall utility operating costs ranged from 2% to 35% with an average of 11%. The City of Canandaigua, New York, estimated that its electrical energy costs are about 27 percent of its water treatment plant budget (Reis, 1999).

The relative distribution of power consumption among process areas for a typical surface water treatment plant is indicated on Figure 5.1. Finished water pumping, raw water pumping, in-plant pumping, rapid mixing, and filtration, generally the most energy intensive processes provide the most potential benefits from energy saving measures.

Table 5.2
Estimated Staffing Requirements
(Full-time equivalents)

	50 mgd (190 m^3/day)		5 mgd (19 m^3/day)	
Position	Semiautomatic	Fully automatic	Semiautomatic	Fully automatic
Operator	15	5	5	1
Instrument Technician	2	3	0	1

Source: Water Treatment Plant Design, Fourth Edition (Adapted from Table 25)

Savings may be realized by implementing energy management strategies such as off-peak pumping, as well as by making operational changes, such as using energy-efficient pumps, instead of less efficient standby pumps, whenever possible.

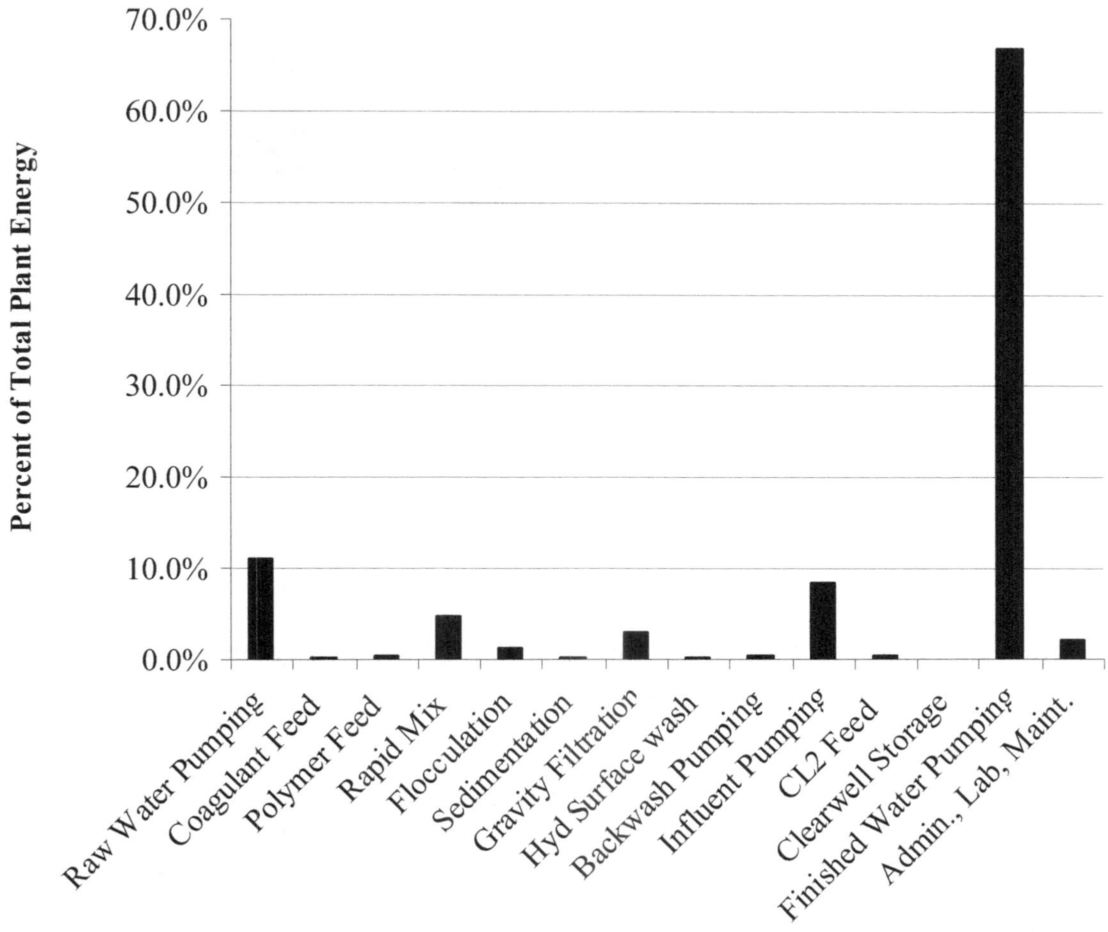

Source: EPRI 1994.

Figure 5.1 - Typical surface water treatment plant energy usage

It should be noted that, although some process areas may have lower total energy consumption, Figure 5.1 does not show peak requirements. As a result, certain process equipment may consume higher amounts of energy for short periods, which can affect the demand charges that the utility incurs.

EPRI's *Water and Wastewater Industries: Characteristics and Energy Management Opportunities,* includes an analysis of electricity requirements for both surface water treatment plants and groundwater pumping systems. The analysis shows that the percentage of total plant energy use by process areas is relatively independent of plant size for plants with capacity ranging from 1 mgd to 100 mgd. The analysis for the groundwater pumping systems shows that over 99% of energy use was for pumping, regardless of system size.

Ozonation, although not covered above, can account for a significant portion (20% or more) of a plant's total energy use (EPRI 1994). The overall energy consumption will vary with the plant size, processes utilized, and with the effluent pump lift required. Typical ranges of energy consumption, for surface water treatment plants not practicing ozonation are in kilowatt-hours per million gallons treated indicated on Figure 5.2.

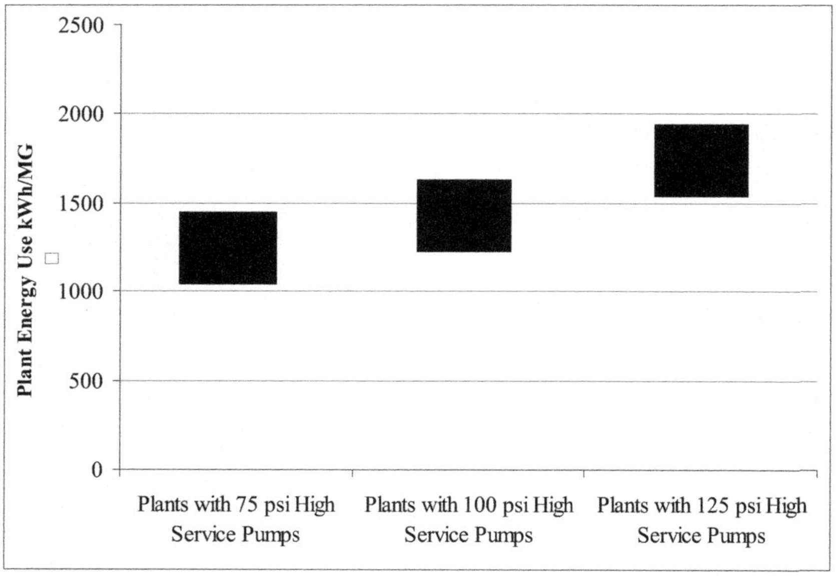

Source: EPRI 1994

Figure 5.2 Ranges of energy consumption for a 10 mgd surface water treatment plant

Chemicals

The U.S. Environmental Protection Agency (EPA) conducted a Community Water System (CWS) Survey in 2000 to obtain data to support development and evaluation of drinking water regulations (USEPA 2000). The results of the survey, the fifth since 1976, incorporated responses from 1,246 water systems. Table 5.3 summarizes some of the data from that study to show many of the common chemical treatment processes used at groundwater and surface water treatment plants. For surface water treatment plants the data indicate that chlorination, coagulation, polymer addition and fluoridation are some of the more common chemical processes.

Table 5.3
Percentage of plants using each treatment process

Treatment Practice	Type of Plant Groundwater	 Surface Water
Chlorination Only	74.3	16.2
Predisinfection/oxidation prior to sedimentation		
Chlorine	7.2	34.5
Potassium permanganate	2.6	20.3

(Continued)

Table 5.3 (continued)

Predisinfection/oxidation prior to filtration		
Chlorine	6.9	23.0
Coagulation/Flocculation	2.1	60.8
Polymers	2.9	46.0
Softening		
Lime/soda ash	6.1	20.5
Post-disinfection after filters		
Chlorine	12.3	68.7
Clearwell	17.7	63.2
Miscellaneous		
Granular activated carbon	0.4	8.6
Aeration	13.4	3.6
Other		
Fluoride	11.4	38.0
PAC	0.0	7.6

Source: US EPA 2000 Community Water System (CWS) Survey, Adapted from tables 23 and 24.

Automation can be used for on-going process adjustments, calculations for process optimization strategies and potentially reduce chemical use. Several examples of potential chemical cost savings are presented in the following sections.

PLANT TYPES AND PROCESSES

The Community Water System (CWS) Survey (USEPA 2000) survey results identified the following processes as being widely used in both surface water and ground water treatment plants of all sizes:

- Coagulation/flocculation/sedimentation
- Dual/multimedia filtration
- Chlorine disinfection

Automation considerations and potential labor, energy and chemical savings are discussed in this chapter. Although the use of membrane processes has increased significantly in the recent years, the plants that use membrane systems represent a low percentage of all water treatment facilities. It is worthwhile to note that membrane plants are commonly highly automated and the level of automation has the potential to support unattended operation. It is noteworthy that corrosion control and fluoridation were also indicated as widely used processes but are not discussed in this chapter.

The following additional processes were not specifically listed in the survey results but are discussed in this chapter due to their potentially significant energy usage, cost-benefits as well as general widespread use.

- Raw water pumping
- Finished water pumping

REPRESENTATIVE WTP PROCESSES

This section presents a review of the automation considerations for several of the common water treatment processes. Figure 5.3 is a simplified schematic of a typical water treatment plant.

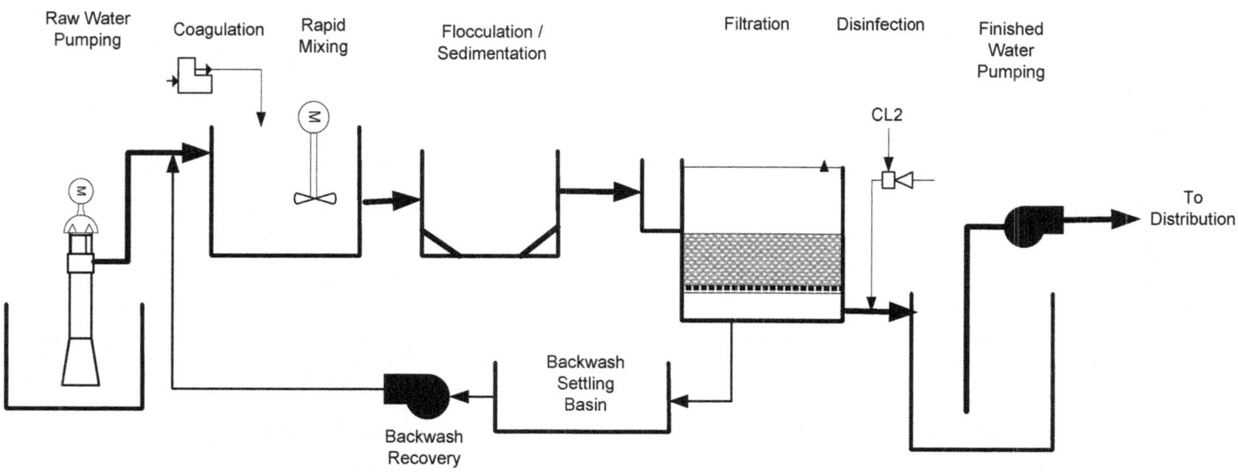

Figure 5.3 Simplified WTP schematic

Raw Water Pumping

Raw water pumping involves moving raw water from the source to the treatment plant. Surface water treatment plant intakes are typically equipped with screens or rakes that remove large objects from the flow before it enters the pumps that convey it to the head of the treatment plant.

The costs of converting raw water pumping to automatic mode of operation include the automatic controls, valve actuators, variable frequency pump drives, and flow and level monitoring instrumentation. Potential benefits include reduced energy cost as a result of off-peak operation, automatic startup and shutdown in response to emergencies or water demands, reduced labor costs, and improved water quality as a result of more uniform flow rates. Figure 5.4 provides a simplified schematic of typical raw water controls.

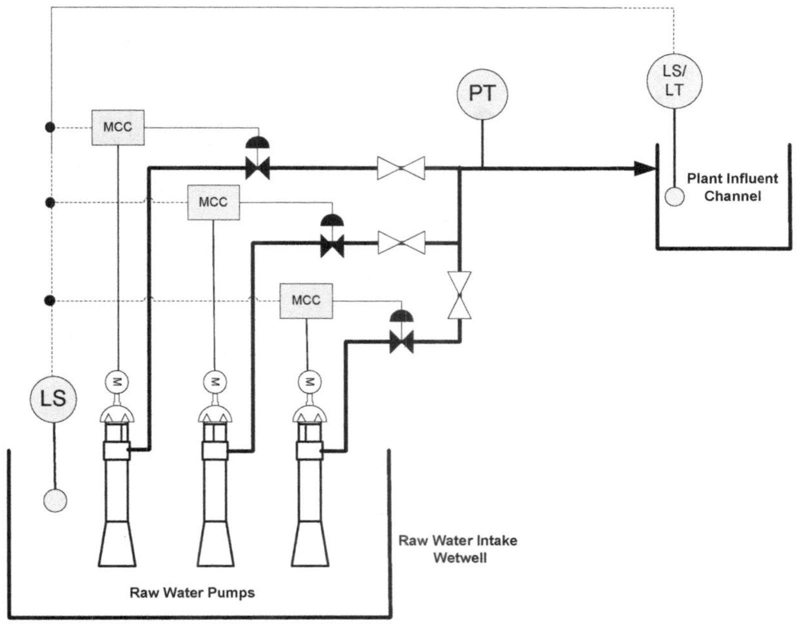

Figure 5.4 Simplified raw water pump control

Levels of Control

Manual. A common method of controlling raw water pumping is by manually adjusting the number of pumps in operation. The plant flow rate is adjusted to match the amount of raw water being delivered to the plant.

Typical Instrumentation:

- Raw water flow meters
- Influent channel level sensor
- Raw water wetwell level sensor

Automatic Control. Automatic control can include adjusting raw water flow to match it to the desired plant flow rate. Figure 5.5 provides a simplified schematic of automatic raw water pumping controls and typical instrumentation includes:

- Raw water flow meters
- Influent channel level sensor
- Raw water wetwell Level sensor
- Flow controller

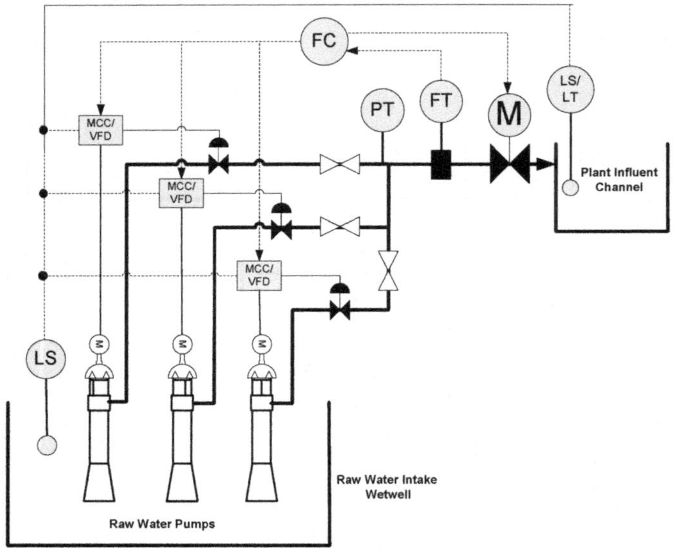

Figure 5.5 Automated raw water flow control

Unattended Operation. Unattended operation in either manual or automatic mode is possible, depending on the complexity of the plant, and the degree of automation. Some automation is typically included downstream from raw water pumping to accommodate variations in plant flow.

Automatic starting and stopping of the raw water pumps combined with overall plant startup or shutdown may be required depending on whether or not the WTP would go on and off line during unattended operation.

Costs

Converting a manually operated plant to automatic operation involves the following additional equipment:

- Speed and valve controls
- Valves and actuators
- Variable frequency drives

Theses devices typically require only periodic maintenance and generally are well proven and reliable.

Potential Benefits of Automation

The benefits of raw water flow control can include more consistent treated water flow potentially having impacts or improvement on water quality. Lower energy costs may result if variable frequency drives are utilized since the pump speed and capacity can be matched to a more uniform flow rate.

Labor Saving. Less travel time to/from the plant to check on pump operation.

Energy Savings. As indicated in Figure 5.1 raw water pumping constitutes a significant percentage of the total water treatment plant energy consumption and depending on the pump capacity, can contribute to utility demand charges. Automation can be used to minimize demand charges and to adjust raw water flow rates to match plant production rate to raw water flow.

- The Madera Valley Water Company installed VFDs and PLCs at two wells, saving $17,000 per year, or 13% of its electricity bill (CEC 2003d).
- The City of Canandaigua, New York, estimated that its electrical energy costs, about 27 percent of its water treatment plant budget are attributable primarily to raw water and finished water pumping (Reis, 1999).
- El Dorado Irrigation District, Placerville, CA. Realized an energy cost savings by shifting raw water pumping to off-peak hours on plant energy use (Chaudhry, 2005).

Risks and Mitigation

Failure of the level sensors or flow meters may result in a mismatch between the amount of water delivered to the plant and could result in an overflow of the influent channel. Many plants are designed with overflow structures that would divert the flow back to the raw water source to avoid any flooding or operational problems.

Table 5.4 identifies general potential risks that may not apply to all plants, and is not intended as comprehensive for all plant situations. Risks should be considered individually for each plant. Potential mitigation strategies for these failures are discussed below under unattended operation.

To mitigate risks during unattended operation, operator alarms or automatic plant flow reductions/shutdowns may be considered based on the following conditions:

- Low raw water wetwell level, redundant low level sensors
- High or low influent channel level sensors and alarms
- High raw water flow rate
- Raw water flowmeter failure

Table 5.4
Potential risks for raw water pumping unattended operation

Device/Item	Risk
Raw water wetwell level sensor	Failure of the low level interlock could result in equipment damage, loss of prime or increased maintenance due to clogging.
Raw water influent channel flow sensor	Instrument failure may not have a direct impact on the control of the raw water pumps but could potentially lead to overflow of influent channel.
Raw water influent channel level	Failure of the high influent level interlock could result in channel overflow, potential equipment damage due to flooding, wasted energy, loss of water.

Redundancy of pumping equipment with automatic sequencing can also be utilized to mitigate the risks of some of the potential failures.

Summary

Unattended raw water pumping operation is possible without the need for automatic control of raw water flow rate. This can be accomplished by starting and stopping the desired number of pumps in order to meet the desired plant flow rate. Other plants will require flow control or level control based on the desired plant production or to maintain constant flow or level control.

Proper sequencing of pumps, using speed control to make gradual changes, or reduction in pump speeds may result in reduced power costs. Simply automating the time of day when the raw water pumps are operated can result in savings due to off peak pumping. Costs can include instrumentation costs, increased calibration and maintenance labor costs, and potential increased influent channel overflows.

Little information was found on applications of advanced control for raw water pumping. A benefit that could be quantified includes improvements in water quality.

Coagulation/Flocculation/Sedimentation

Coagulation involves the addition of chemicals to influent raw water to form particles large enough to be removed by settling. Typical coagulants are alum, ferric chloride, and coagulant aids. As raw water pH can interfere with the coagulation process, pH control is often included.

Control Modes

Manual Control. A common method of controlling coagulant dosages is manually adjusting the coagulant, coagulant aid, and acid or caustic feed rates based upon observation, jar tests, and instrument readings. Jar testing is performed on a scheduled basis; with increased frequency when the quality of source water fluctuates. Figure 5.6 provides a simplified schematic of coagulation controls with limited automation and typical instrumentation includes:

- Raw water flowmeters
- Raw water turbidity monitors
- Raw water pH meters
- Settled water turbidity monitors

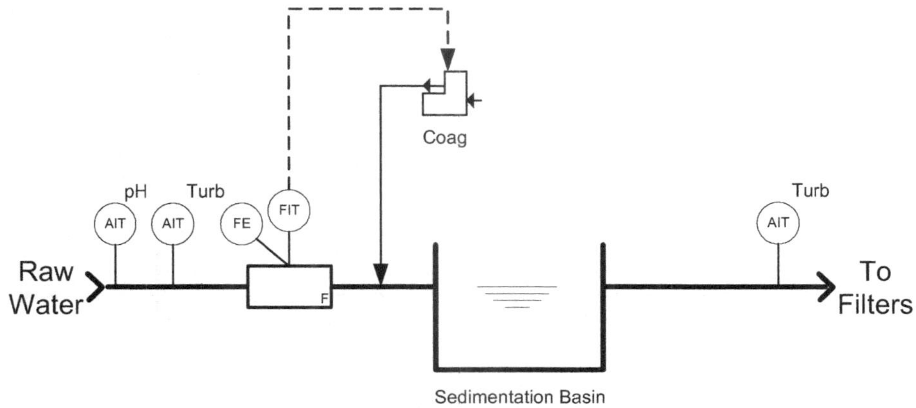

Figure 5.6: Example coagulation control with minimal automatic control

Automatic Control

Automatic control adjusts coagulant feed rates and dosages in response to variations in plant flow and/or source water quality. Automatic shutdown of the plant should also shut down coagulant feed. Automatic pH control may be required at some plants. (Dentel and Kingery 1988). Feedback control of the coagulant dose is accomplished using streaming current detectors and controllers. The use of streaming current detectors may not be suitable with some source waters. Figure 5.7 provides a simplified schematic of coagulation controls with automatic controls and typical instrumentation includes:

:

- Raw water flowmeters
- Raw water turbidity monitors
- Raw water pH meters
- Settled water turbidity monitors
- Settled water pH meters*
- Controller for streaming current detectors*
- pH controllers*

* Where required

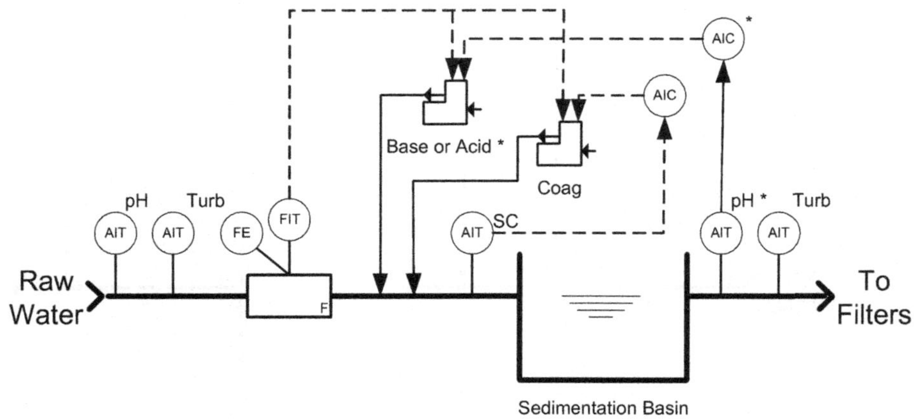

Figure 5.7 Example automated coagulation control

Unattended Operation

Unattended operation in manual mode is possible with appropriate provisions for the variations in source water quality, such as regular adjustment of chemical dosages. Some automation is typically included to adjust for plant flow variations.

An example of a plant that operates partially unattended is the Umpqua Water Treatment Plant (Groshong 2006), which draws raw water from the North Umpqua River. The plant's coagulant metering pump speed is automatically adjusted based on influent flow rate, and the pump stroke is manually adjusted by the operator based on the raw water turbidity and jar test results.

The plant operates unattended for five to nine months, typically from April until November, depending on the weather. During this period, jar testing is conducted twice daily and chemical doses are adjusted as needed. Raw water turbidity during this period is 0.8 to 3 NTU, and the jar tests take about 15 to 30 minutes, depending on water temperature. During the winter the plant operates in an attended manner. The jar testing frequency is adjusted according to weather and river conditions, and during high flow and high raw water turbidity (up to 500 NTU) testing may be done every 30 to 45 minutes.

Costs

The additional equipment needed to convert a manually operated coagulant feed process to automatic control typically includes the following items:

- Controller – coagulant dosage
- Streaming current detector*
- Controller – pH*
- Settled water pH monitor*
- pH controller*
- Chemical feed system for pH control*
  * Where required

Streaming current detectors require regular cleaning. The frequency of cleaning may vary from every 2 days to every 3 months depending on influent water quality and the type of coagulant used (Dentel and Kingery 1988).

Benefits

The benefits of adding streaming current detectors can include improved coagulant dosing resulting in chemical cost savings, more consistent treated water quality, and fewer process upsets (Hargesheimer, Conio and Popovicova 2002).

Rapid mixers provide energy for mixing the chemicals and promoting floc formation. Energy savings may be available by optimizing floc formation and chemical mixing; using VFDs on this equipment will allow operator control to achieve this optimization.

- The San Juan Water District estimates it gained annual savings of $11,000 as well as improved process control by installing four VFDs to control ten, 5 hp flocculation pump motors and six small chemical feed pumps at its 120 mgd Peterson Water Treatment Plant. This savings represented approximately 30% of the flocculation energy costs, and an undetermined amount of chemical feed energy costs (CEC 2003c).

Chemical Cost Savings and Improved Water Quality. The most significant study pertaining to the automation of coagulant dosing identified was *An Evaluation of Streaming Current Detectors* (Dentel and Kingery 1988), which evaluated and showed numerical results for ten water treatment plants that have implemented automatic coagulant control using a streaming current detector. Of the ten plants, two changed coagulant chemicals when the streaming current detectors were installed, and were therefore, not included in the overall cost results. Chemical cost results from the study are listed in Table 5.5.

When raw water conditions were stable, plant chemical use per MG decreased by up to 45%, with an average reduction of 12%. Chemical use at two of the plants showed a slight increase. During periods of variable raw water conditions, chemical usage per MG decreased in all plants by up to 63%, with an average of 23%.

The study compared turbidity levels before and after the installation of the streaming current detector to evaluate its impacts on water quality. Overall results showed a decrease in turbidity in four plants, an increase in two plants, and little change in the remaining four plants. These tests focused only on overall turbidity, not on changes in consistency of water quality. The average plant flows increased by 21% between the 'before' tests and the 'after' tests. The reason for the increase is unknown, but it may affect interpretation of the results. Statistically, this data indicated no significant difference at the 95 percent confidence level.

The Dentel study included a survey that tabulated the results from 35 utilities that use streaming current detectors for either information or closed-loop control. Of the 35 utilities, 94% had experienced fluctuations of 1 to 2 orders of magnitude in raw water turbidity. In response to the survey question about chemical use and water quality, 23% of the utilities reported savings in chemical use and 25% reported improved ability to respond to transient conditions and/or equipment malfunction. A number of utilities also report improved water quality consistency and filter run times. The survey results are in Table 5.6. (Note: nine of the utilities marked more than one response).

Table 5.5
Cost and payback period analysis before and after SCD installation

Plant	1	2	4	5	6	7	9	10
Cost	$ 7,500	$ 9,300	$ 26,400	$ 6,000	$ 10,400	$ 8,800	$ 7,500	$ 6,000
Installation	---	---	$ 300	$ 70	$ 900	$ 120	$ 200	---
Repair Cost	---	---	$ 1,090	---	---	---	---	---
STABLE CONDITIONS								
Flow (MGD)								
Before SCD	4.3	6.8	24.7	23.4	10.0	18.9	3.5	
After SCD	5.7	7.8	30.7	30.5	10.0	22.8	4.6	
Chemical Savings								
(Per MGD)	$ 2.10	$ (0.90)	$ (0.20)	$ 1.20	$ 1.70	$ 0.20	$ 1.20	
(Percent)	19.5	-3.4	-3.8	12.0	45.0	4.4	7.0	
Payback Period (Years)	1.7	-4.1	-15.8	0.5	1.1	6.0	1.6	
CHANGING CONDITIONS								
Flow (MGD)								
Before SCD		7.6	24.6	21.0	8.0	20.8	8.5	9.9
After SCD		8.7	30.8	26.6	8.0	19.4	10.9	8.5
Chemical Savings								
(Per MGD)		$ 0.90	$ 0.30	$ 3.30	$ 10.60	$ 0.50	$ 4.30	$ 12.20
(Percent)		2.9	6.0	24.0	63.0	10.0	20.0	33.0
Payback Period (Years)		3.4	8.7	0.2	3	2.9	0.5	0.1

Source: Dental and Kingery (1988)

Table 5.6
Utility survey of streaming current detector effects

Comment	Frequency
EFFECT ON CHEMICAL USE	
Reduction in chemical usage	8
Little to no effect on chemical usage	3
No effect on chemical usage	3
Chemical usage has not been evaluated	4
EFFECT ON WATER QUALITY	
Better response to transient conditions and/or equipment malfunction	9
Better and/or consistent water quality	6
Better setting floc and/or improved filter run time	4
Other	2
No response/No comments	6

Source: Dental and Kingery (1988)

In 1989, a survey of 26 water treatment plants operated by American Water Works Service Company that utilized streaming current detectors reported that 38% of the plants experienced a decrease in chemical costs attributable to use of these devices; 54% of the plants experienced better finished water quality; and several noted longer filter runs.

Labor Savings. Labor savings can include time spent on jar testing (although some jar testing would still be performed, its frequency can be reduced), and the travel time to/from the plant to check status (for partially attended plants).

As described for the Umpqua Water Treatment Plant, during periods of rapidly changing water quality, an operator spends significant time monitoring and adjusting the chemical dosage. Noel Groshong, the General Manager of the Umpqua Basin Water Association, estimates that the average time spent on jar testing and controlling coagulant dosages at two person-hours per day in the summer including travel time, and an average of nine person-hours per day in the winter during continuously attended operation.

Risks and Mitigation Strategies

Manual Mode. Failure of turbidimeters or pH analyzers will not cause a process upset, since there is no closed-loop control from these instruments. Table 5.7 identifies general risks that may not apply to all plants, and is not intended to include all plant situations. Risks should be considered individually for each plant. Possible mitigation strategies for such failures are discussed below under unattended operation.

Automatic Mode. Significant changes in raw water temperature, color, turbidity, or pH can cause the streaming current setpoints to become less accurate (Dental and Kingery 1988). Note that streaming current detectors may react differently at different plants because of different processes and variations in raw water quality.

Table 5.7
Manual mode, general risks

Device/Item	Risk
Raw water turbidity	Sudden changes resulting in poor coagulated water quality leading to declining production and possibly poor finished water quality
Raw water pH	Sudden changes resulting in poor coagulated water quality leading to declining production and possibly poor finished water quality
Raw water – color, organics, alkalinity	Sudden changes potentially resulting in poor finished water quality
Flowmeter	Instrument failure resulting in poor coagulated water quality leading to declining production and possibly poor finished water quality
Coagulant/aid feed system	Failure or loss of chemical feed resulting in poor coagulated water quality leading to declining production and possibly poor finished water quality

Pilot testing of streaming current analyzers may be beneficial before permanent installation to confirm streaming current set point values and suitability of the technology.

Risks should be considered individually for each plant. To mitigate the risks during unattended operation, operator alarms or automatic plant flow reductions/shutdowns may be considered under the following conditions:

- High raw water turbidity
- High settled water turbidity
- High effluent turbidity
- High/low raw water pH
- High/low streaming current
- Streaming current detector failure*
- Raw water flowmeter failure*
- pH transmitter failure*
- Loss of coagulant flow (flowmeter or flow switch)
- High/low pilot plant settled water turbidity
- High/low pilot plant effluent turbidity

* When used for closed-loop control

Table 5.8 lists general risks that may not apply to all plants, and is not intended to include all plant situations. Redundant equipment and instrumentation can be used to mitigate the risk of failures.

Table 5.8
Automatic mode general risks

Device/Item	Risk
Raw water turbidity	Sudden changes resulting in poor coagulated water quality leading to declining production and possibly poor effluent water quality
Raw water pH	Sudden changes resulting in poor coagulated water quality leading to declining production and possibly poor effluent water quality
Raw water – color, organics, alkalinity	Sudden changes potentially resulting in poor effluent water quality
Flowmeter	Instrument failure resulting in poor coagulated water quality leading to declining production and possibly poor effluent water quality
Coagulant/aid feed system	Failure or loss of chemical feed resulting in poor coagulated water quality leading to declining production and possibly poor effluent water quality
pH transmitter/ controller	Instrument failure resulting in poor coagulated water quality leading to declining production and possibly poor effluent water quality
Streaming current detector/ controller	Instrument failure resulting in poor coagulated water quality leading to declining production and possibly poor effluent water quality

Advanced Control

Neural networks for optimization of coagulation dosages have been studied and used in both open-loop and closed-loop control applications (Hargesheimer, Conio and Popovicova 2002; Baxter et al. 2001). They can also be used as an alternative and/or an enhancement to streaming current detectors. The use of neural networks for control is not widely practiced at present; however, there is growing interest and consideration of their use.

The use of on-line particle counters for optimization of coagulant dosing is described in the publications *Operational Control of Coagulation and Filtration Processes* (AWWA Manual M37) and *Online Monitoring for Drinking Water Utilities*. Although these systems are on-line, they do not include closed-loop control of the process. The implementation and use of on-line particle counters to manually optimize coagulant dosing at the Alfred Merritt Smith Water Treatment Facility (direct-filtration plant) of the Southern Nevada Water System resulted in a 32% reduction in chemical costs, longer filter runs, and a load reduction in the sludge handling system (AWWA M37 2000).

Summary

In some plants, unattended operation is possible without the need for automatic control of coagulant dosing. Other plants will require feed-forward flow control, feed-back control from a streaming current detector, and/or pH control.

Savings in labor and chemical costs, longer filter run times (resulting in reduction in both power and sludge handling costs) and improved water quality are additional benefits that may be realized by automating the coagulant control. Costs can include instrumentation, increased calibration and maintenance labor, and the potential for an increase in effluent turbidity.

Dual/Multimedia Filtration

The majority of surface water treatment plants utilize gravity fed, granular media filtration to remove suspended solids. There are two primary methods for the control of gravity filters: constant-rate and declining-rate. With constant rate filtration, the influent channel water level is maintained at a constant level as flow is split equally between all on-line filters. There are three main approaches used for controlling constant rate filters including: Constant rate with filter effluent rate of flow controller; constant rate with constant water level and influent flow splitting; constant rate with variable water level and influent flow splitting.

Declining rate filters are typically equipped with effluent weirs or fixed position effluent valves rather than effluent rate of flow controllers. As declining rate filters collect solids, the flow through the filter begins to drop off.

As solids begin to accumulate in the filter media bed, porosity decreases and the head-loss across the bed increases. To avoid head-loss increases beyond a desired level or turbidity break through of the filter media, filters need to be cleaned or backwashed. There are four basic backwash methods including:

- Upflow of washwater through the filter bed without auxiliary scour
- Upflow of washwater through the filter bed with air scour

- Upflow of washwater with surface wash
- Continuous back wash

The method of backwashing depends on the plant specific requirements. Monitoring and control of filter effluent turbidity is a primary regulatory requirement. The Enhanced Surface Water Treatment Rule (ESWTR) is the primary regulation that applies to the performance of filters and establishes maximum permitted filter effluent turbidity requirements. The requirements for turbidity measurements and methods are defined in the USEPA Long Term 1 Enhanced Surface Water Treatment Rule (LT1ESWTR) Turbidity Provisions *Technical Guidance Manual* (EPA 816-R-04-007, August 2004).

Constant-Rate Control

There are three generally recognized ways to operate a filter in constant rate mode:

- Utilizing a rate of flow controller in the filtered water piping
- Influent flow splitting to multiple parallel filters, with a constant filter level
- Influent flow splitting to multiple parallel filters with variable filter level

Automatic Control. A certain level of automated control is required for filter operation; completely manual control is typically not utilized. The most common indicator of filter performance is turbidity. Each filter's effluent turbidity should be monitored and recorded continuously. Figure 5.8 provides a simplified schematic of a typical filter configuration and typical instrumentation includes:

- Individual filter turbidity
- Combined filter effluent turbidity
- Filter effluent flow rate
- Filter influent channel level
- Filter headloss
- Filter level or level switches
- Backwash flow rate
- Filter effluent particle counts
- Flow controller – using modulated discharge control valves to vary filter effluent flow rate

Backwash Control

Backwash is typically initiated manually with operator intervention or automatically based on filter headloss, filter run time or turbidity levels of the filtered water exceeding a certain setpoint. Backwash sequencing typically is configured to minimize filter bumping by gradually increasing backwash flow rate or having a low rate and high rate backwash. Backwashing is typically completed based on the duration of the backwash cycle or on backwash water turbidity.

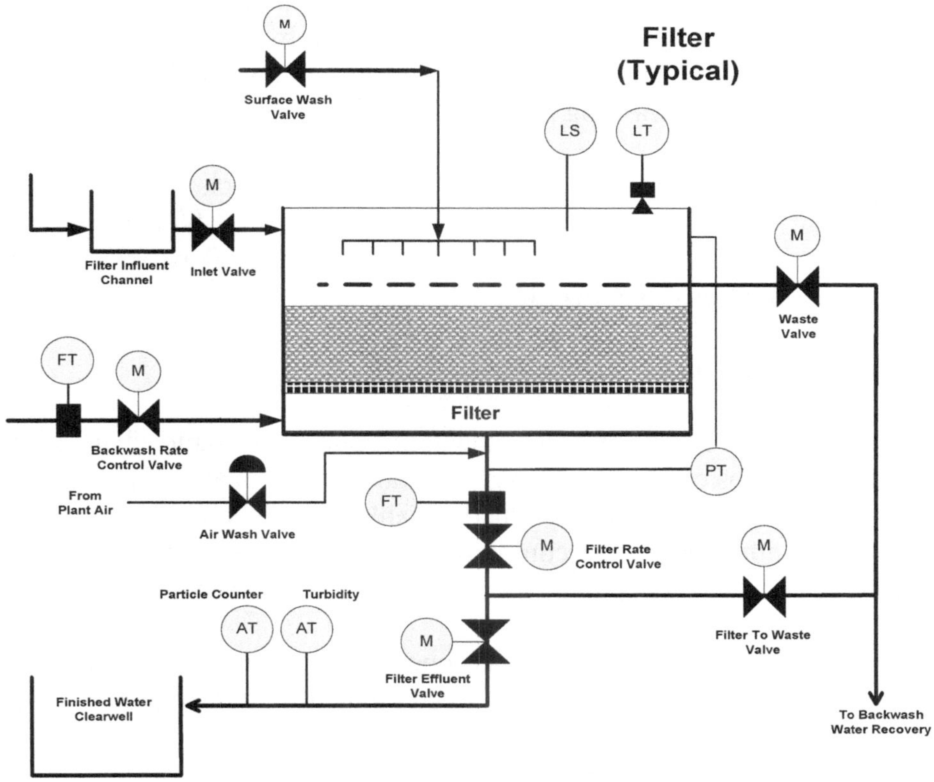

Figure 5.8 Example filter flow control

Benefits

The City of Canandaigua, New York, modified the instrumentation and control system at its WTP to incorporate both particle counters and turbidimeters (Reis, 1999), to establish and operate an optimal backwashing procedure. The automation was essential in implementing this new control approach that included modifications to the backwashing sequence to minimize filter bumping and to start the high rate washing step earlier than with timer and terminate the backwashing run based on turbidity measurements. The benefits of the improvements included the following:

- Finished water productions savings of over 35%, approximately 17 million gallons per year
- Energy savings of 31,200 kWh/year which resulted in annual savings of about $2,200
- Chemical cost savings of about $990 per year
- Operating time savings of about 80 hours per year
- Deferred capital improvements (necessary for expanded plant capacity) equating to about $12,600 per year
- Non-quantifiable benefits included preservation of a natural resource (Canandaigua Lake has limited capacity to serve), improved potable water quality and reduced risk of failure to meet water quality regulations

Although backwash pumps do not consume large amounts of energy, they can contribute significantly to demand charges. Savings can be realized by shifting the backwashing to off-peak hours. Other plant loads, such as the finished water pumps, can be automatically turned off or turned down during the backwashing process to reduce overall energy demand (EPRI 1994).

- By incorporating off-peak pumping at the Hemlock Water Treatment Facility, Aquarion Water Company in Connecticut saves \$9,600 per year, and by implementing off-peak backwashing, it saves \$25,100 per year without adversely affecting operations (Schultz 2003).

Risks and Mitigation Strategies

Automatic Mode. Failure of the effluent turbidimeters, rapid changes in flow rate or exceeding design flow rates can be detrimental to treated water quality. Some of the general risks associated with unattended operation are in the Table 5.9. These risks may not apply to all plants, and the list is not intended as comprehensive for all plant situations. Risks should be considered individually for each plant. Potential mitigation strategies for failures are listed below.

To mitigate the risks during unattended operation, operator alarms or automatic plant or filter flow reductions/shutdowns may be considered on the following conditions:

- Filter level reading out of tolerance
- High filter effluent water turbidity
- High filter effluent flow
- High filter water particle counts

Redundant process equipment and instrumentation can help to mitigate the risk failures.

Table 5.9 Potential mitigation strategies

Device/Item	Risk
Filter effluent water turbidity	Analyzer failure resulting in undetected water quality issues leading to declining production and possibly poor effluent quality
Filter effluent water particle counts	Analyzer failure resulting in improper coagulant dosage and or undetected water quality issues leading to declining production and possibly poor effluent water quality
Filter level	Failure of instrument or level switches could result in improper filter levels when initiating backwashing resulting in poor effluent water quality
Effluent flow meter	Instrument failure resulting in rapid flow changes leading to possibly poor effluent water quality

Summary

Some plants can be operated unattended without the need for automatic control of filter backwashing. Savings in labor and chemical costs, longer filter run times (which result in reductions in power and sludge handling costs) and better water quality are additional benefits that may be realized by automating filter backwashing. Costs can include instrumentation, increased calibration and maintenance labor, and increased effluent turbidities.

Chlorine Disinfection

Several USEPA and Safe Drinking Water Act regulations affect how or when a water system utilizes oxidation or disinfection. Chlorination of filtered water prior to entry into the distribution system is a commonly used disinfection process (USEPA 2000). This section includes automation considerations for chlorine disinfection.

For water systems serving more than 3,300 persons, the SWTR requires continuous monitoring of disinfection residual where the treated water enters the distribution system. The residual disinfection concentration cannot be less than 0.2 mg/L for more than 4 hours. In the event that the residual disinfection goes below 0.2mg/L the state must be notified. The lowest value for each day needs to be reported and the date and duration when disinfection residual was less that 0.2 mg/L and when the state was notified.

Control Modes

Control system design is an important aspect of effective chlorination and typically uses one of the following methods:

- Manual
- Automatic, including
    - Flow paced
    - Residual feedback
    - Compound closed-loop

Manual. Figure 5.9 shows a typical schematic of manual chlorination control. A fixed dosage is manually set so that the chlorinator delivers a constant rate of chlorine while the plant is operating. The dosage setting is typically determined based on the plant flow rate, the chlorine demand and the desired chlorine residual. Adjustment, or fine-tuning, of the dosage rate is made manually based on plant flow readings and periodic or continuous residual readings.

This approach is most effective when plant flow rates and chlorine demand remain relatively constant. If the plant flow rates are not constant, significant swings in the finished water residual can occur. This situation may require plant operator intervention to make manual adjustment of the dosage rate in order to keep residual levels within the desired range. Typical Instrumentation for this control mode includes:

- An interlock to start the chlorinator when plant is operating
- Continuous residual analyzer

- Residual recorder

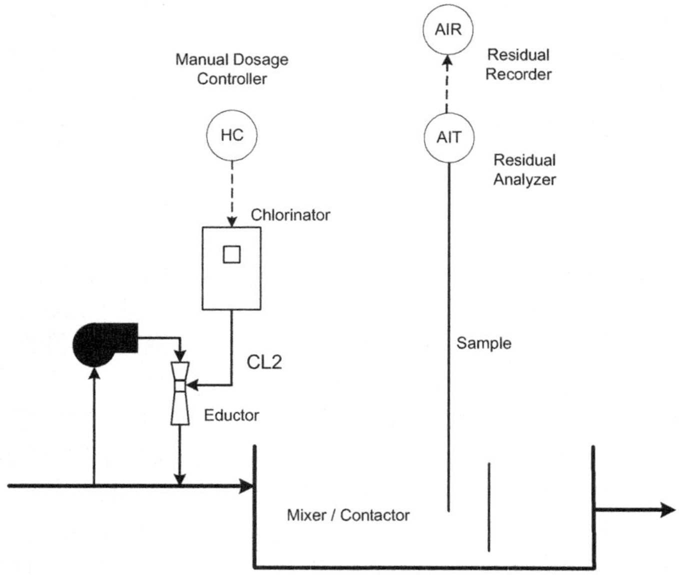

Figure 5.9 Manual chlorination control

Automatic Control. The following approaches, in order of increasing complexity, are used for automatic chlorination control:

- Flow pacing
- Residual feedback control
- Compound closed loop control suing flow and residual feedback

Figure 5.10 provides a schematic of a typical compound closed-loop chlorination control where the dosage is based on a combination of plant flow rate and continuous feedback of the chlorine residual. This approach provides automatic adjustment of the dosage rate to accommodate for changes in plant flow rate and chlorine demand. However, is dependent on the proper operation of the chlorine residual analyzers which do require routine maintenance and calibration and can be subject to signal drift and reliability issues. Typical Instrumentation:

- Filtered water flow
- Residual analyzer
- Residual recorder
- Residual controller – configured for one of the following modes: flow paced, residual control or compound closed-loop control.

Unattended Operation. Unattended operation is possible in both manual and automatic mode. In the manual mode, both plant flow rate and residual demand must be constant for stable residual control.

Costs

The equipment needed to add automatic control typically includes the following:

- Residual controller
- Chlorinator with pacing signal input

Theses devices typically require only periodic maintenance and are generally reliable.

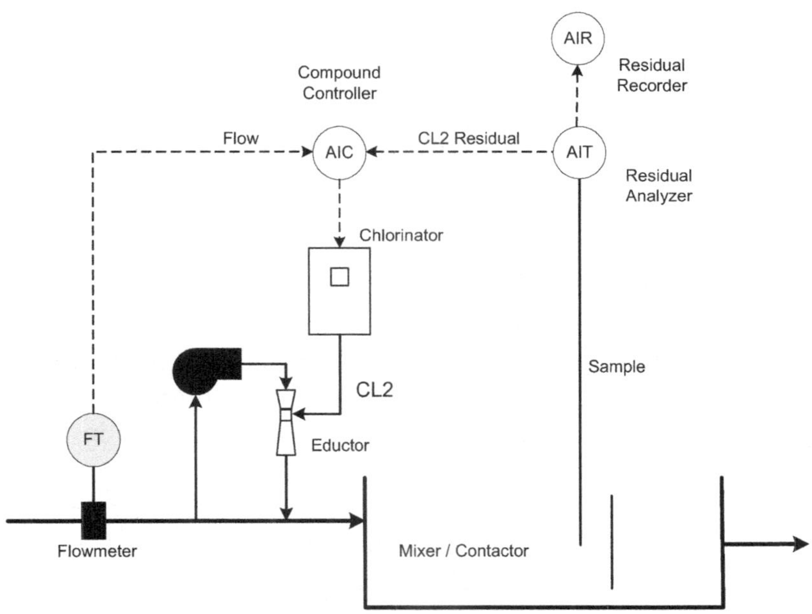

Figure 5.10 Automatic chlorination control

Potential Automation Benefits

Chemical Savings. Benefits of chlorination automation can include more consistent treated water chlorine residual, potentially lower chemical costs since the dosage control can be used to avoid over feeding of chemicals.

Labor Savings. The reduction in the labor for process adjustments and travel time to/from the plant.

Energy Savings. The energy costs associated with operating chlorine disinfection systems are relatively low and this is not considered a significant area for energy savings.

Risks and Mitigation Strategies

Automatic Mode. Failure of the chlorination controls can have a detrimental effect on finished water quality. Failure of chlorine residual analyzers in the finished water storage reservoirs could have an effect on water quality. Inadequate chlorine dosage could result in too low of a residual. Too high of a pumping rate with inadequate chlorine residual could result in not achieving the required CT time of the finished water.

Some of the general risks and potential mitigation strategies for failures associated with unattended operation are in Table 5.10. These risks may not apply to all plants, and the list is not intended as comprehensive for all plant situations. Risks should be considered individually for each plant.

To mitigate the risks during unattended operation consider the following:

- High and low chlorine residual alarms

Redundant process equipment and instrumentation can help to mitigate the risk failures.

Summary

Some plants can be operated unattended without the need for automatic control of chlorine dosage. However, feed forward flow control appears to be the minimum preferred approach when considering unattended operation. Residual control can theoretically provide more precise control when flow rates and changing chlorine demand occur. Analytical equipment requires on going maintenance and calibration. Savings in labor and chemical costs may be realized by automatic control of chlorine residual.

Table 5.10
Potential mitigation strategies

Device/Item	Risk
Finished water flow rate	Flow meter failure could result in improper flow pacing control and chlorine residuals either too high or too low.
Finished water chlorine residual analyzer	Analyzer failure resulting in improper chlorine residual. If undetected could lead to water quality issues, possibly poor effluent water quality or failure to meet required CT. Readings below 0.2 mg/L require reporting to State. Routine maintenance and calibration Consider backup analyzer

Finished Water Pumping

Finished water pumping involves moving treated water from the plant clearwell into the distribution system using high service pumps. Considerations associated with unattended operation include automatic starting and stopping of the high service pumps, variable speed controls for flow, level, or pressure; and safety interlocks for shutdown on low inlet water or high discharge level. Constant speed pumping may include automatic alternation of pumps or sequencing combined with discharge valve control.

Costs include providing automatic level/pressure/flow controls, valve actuators, variable frequency drives and flow and level monitoring instrumentation. Benefits include savings in energy cost through off-peak operations, reduced labor costs, automatic startup and shutdown in response to emergencies or water demands, and improved distribution system pressure control.

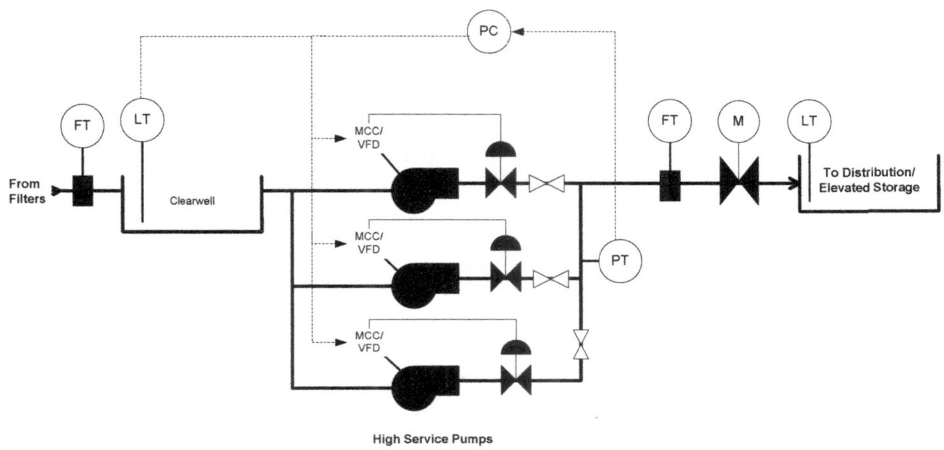

Figure 5.11 Simplified high service pump controls

Control Modes

Manual. A common method of controlling finished water pumping is by selecting the proper number of fixed-speed pumps to match plant production. Finished water is typically moved to the distribution system or to reservoirs, and plant flow rate is adjusted to match the demand. Typical Instrumentation:

- Finished water flow
- Clearwell level
- Finished water reservoir level
- Finished water chlorine residual
- Distribution system pressure

Automatic Control. Automatic controls can adjust the finished water flow rate to match system demand. This can be done based on reservoir level, distribution system pressure, or a preset flow rate. Figure 5.11 provides a simplified schematic of hig service pump controls and includes the following typical instrumentation:

- Finished water flow
- Clearwell level
- Finished water reservoir level
- Finished water chlorine residual
- Distribution system pressure
- Flow controller – using modulated pump discharge control valves or adjustment of pump speed using variable frequency drives.

Unattended Operation. Unattended operation is possible in both manual and automatic mode, depending on the complexity of the plant, and the ability of plant automation to adjust to fixed or variable finished water flow rate.

Costs

The equipment needed to add automatic control typically includes the following:

- Speed or valve controls
- Valves and actuators*
- Variable frequency drives*

  * These may or may not be included in the economic analysis.

Theses devices typically require only periodic maintenance and are generally reliable.

Potential Automation Benefits

The benefits of finished water pump control can include more consistent treated water flow, lower energy costs where variable frequency drives are utilized since the pump speed and capacity can be matched to a more uniform flow rate and, where system storage is available, savings in energy costs through off-peak pumping.

Labor Savings. The reduction in the labor for process adjustments and travel time to/from the plant.

Energy Savings. Finished water pumping can represent a substantial percentage of the plant energy consumption, and depending on the pump size, can contribute significantly to utility demand charges. Automation can be used to minimize demand charges and to adjust raw water flow rates to match the plant production rate with system demands. Where system storage is available, automation can be used to shift production to off peak periods. For example:

- The East Bay Municipal Utilities District implemented an on-line optimization program which scheduled its water delivery for a portion of the distribution system, saving approximately $300,000 in its first 11 months of operation (AwwaRF 2005).

- Moulton Niguel Water District reduced its energy bill by nearly $320,000 per year by using PLCs to control off-peak pumping at 77 pump stations in the distribution system. This represents a reduction of over 20% in energy costs (CEC 2003b).
- By incorporating off-peak pumping at the Hemlock Water Treatment Facility, Aquarion Water Company of Connecticut saved $9,600 per year. (Schultz 2003).

Risks and Mitigation Strategies

Automatic Mode. Failure of the Clearwell level sensors can have a detrimental effect on the high service pumps if the level is reduced below acceptable levels.

Failures of the finished water storage tank level sensors could have the effect of not maintaining adequate storage in the reservoirs or if pumping is not shut off on high level the result could be overflow of the reservoirs. Failure of chlorine residual analyzers in the finished water storage reservoirs could have an effect on water quality. Inadequate chlorine dosage could result in too low of a residual. Too high of a pumping rate with inadequate chlorine residual could result in not achieving the required CT time of the finished water.

Some of the general risks and potential mitigation strategies for failures associated with unattended operation are in Table 5.11. These risks may not apply to all plants, and the list is not intended as comprehensive for all plant situations. Risks should be considered individually for each plant.

To mitigate the risks during unattended operation, operator alarms or automatic plant or filter flow reductions/shutdowns may be considered on the following conditions:

- Clearwell low level and low low level alarms
- High finished water reservoir level
- Low finished water reservoir level
- High and low chlorine residual alarms

Redundant process equipment and instrumentation can help to mitigate the risk failures.

Table 5.11
Potential mitigation strategies

Device/Item	Risk
Clearwell level sensor	Sensor failure could result in draining of the clear well as a result of failure to shut off finished water pumps. If suction head on the pumps is too low it could result in pump damage. Failure could also result in frequent pump cycling.
Finished water reservoir level sensor	Failure of instrument or level switches could result in improper storage levels. Either inadequate storage or overflow.
Finished water chlorine residual analyzer	Analyzer failure resulting in improper chlorine residual and if undetected could lead to water quality issues, possibly poor effluent water quality or failure to meet required CT

Summary

Some plants can be operated unattended without the need for automatic control of finished water pumping. Savings in labor and energy costs may be realized by automatic control of pumping particularly if pumping can be done off-peak. Finished water quality could be impacted by automation failures. Pump damage can occur if pumps suction water levels are reduced below acceptable levels or if automation failures result in frequent pump cycling.

ADDITIONAL ENERGY CONSIDERATIONS

This section provides additional WTP energy considerations. It is not the intent to imply that all of the potential savings described in this section will be the result of operating a plant in an unattended manner. However, advanced automation needed for unattended operation may have a secondary benefit if it can enable energy cost saving strategies such as peak shaving and off-peak production and pumping.

Energy Rates

Electric utilities often offer several types of rate structures, from flat rates, to rates that vary based on time of use. Typically, electrical energy rates consist of several separate charges. Those most significant to automation are energy charges and demand charges based on time-of-use.

Energy Charges

The energy charge is the rate per kWh for the energy consumed. Many rate schedules include different rates for different times of the day. Rates can vary by season, day of week, and holidays. Figure 5.12 shows example energy rates for time of use schedule. If automation can shift energy use to periods of lower rates, cost savings should result. Often, the lowest energy rates coincide with periods best suited for unattended operation. Thus, automation to allow unattended operation may enable plants to operate during periods of lowest energy cost.

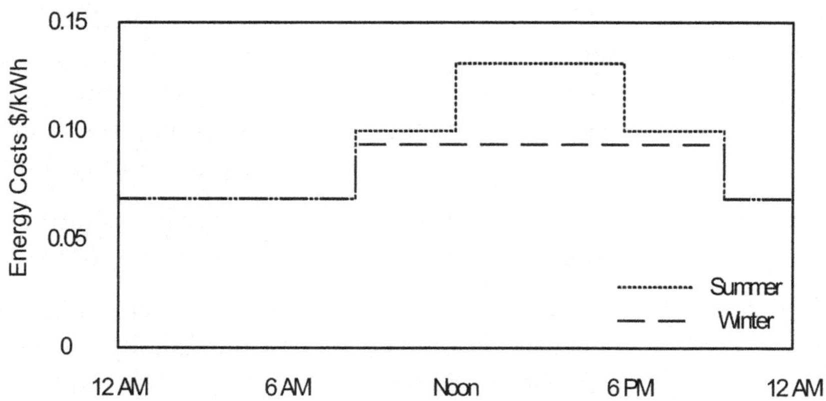

Source: Adapted From Pacific Gas & Electric E-20 Rate Schedule, 2006.
Figure 5.12 Example energy rates for time of use schedule

If shifting plant operation to unattended periods also results in energy saving, the saving could be attributed to the unattended operation. Since pumping is the most significant energy consumer, it would be important to reduce plant flow during peak energy use periods. This approach would involve consideration of distribution system storage, customer demands, and water aging.

For equipment that is operated occasionally, automation can be used to schedule its use to off-hours or sequence its operation of multiple units to reduce demand charges. Rates vary among electric companies, as well as among customers' facilities. Some electric utilities provide real-time pricing rates that change daily. Again, a higher level of automation may allow use of more sophisticated methods of energy management. Examples of off-peak savings are given below:

- The Erie County Water Authority modified the operation of the Sturgeon Point water treatment plant to treat 35 mgd during peak periods and 55 mgd during off-peak hours, which resulted in annual savings of $50,000 (Porter 1996).
- Southeast Water/Wastewater Utility, which uses a real-time pricing rate offered by the electric utility, adjusts its power based upon the real-time rates it locks in on the previous day (Jacob, Kerestes, and Riddle 2003). Therefore, a strategy that reduces energy use during higher cost periods and shifting it to lower cost periods could translate into cost savings.

Demand Charges

Demand charges are generally determined by taking the highest average kilowatts used over a 15 or 30 minute period for the demand period. Demand periods can run from one month to a year. Thus, the utility can be penalized for the highest 30 minute demand charge for up to a year. Demand charges can account for a significant amount of a utility's energy bill. Figure 5.13 shows example demand rates for time of use schedule.

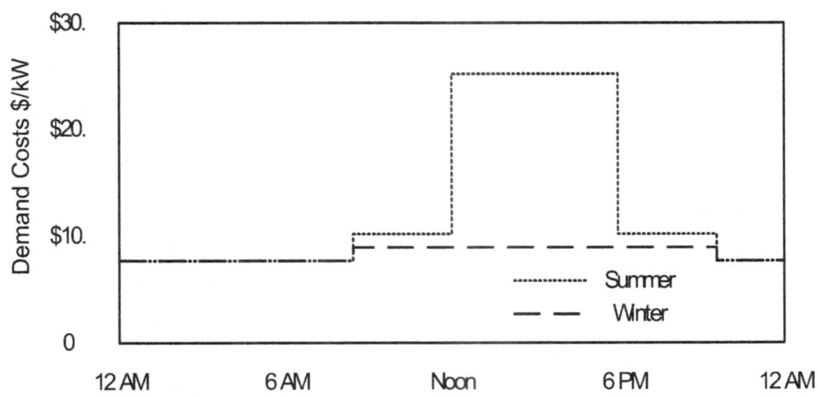

Source: Adapted From Pacific Gas & Electric E-20 Rate Schedule, 2006.
Figure 5.13 Example demand rates for time of use schedule

The following examples illustrate how shifting the time of operation of some equipment has resulted in cost savings through reducing peak demand:

- EPRI's *Water and Wastewater Industries: Characteristics and Energy Management Opportunities* includes an analysis of shifting several loads at a generic 10 mgd surface water treatment plant. The analysis shifts daily backwashing from 10 AM to 8 AM, shifts the washwater decant pumping from 1 hour after backwashing to the afternoon when peak demand subsides, and shifts the residuals pumping to earlier in the morning. In this example, treatment plant flows were not shifted, but follow a typical demand curve. This relatively minor shift in operations resulted in a peak demand reduction of 20% (EPRI 1996).
- American Water analyzed energy use at the following water treatment plants: San Katy WTP, the Illinois River WTP, the Hays Mine WTP, and the Milton/White Deer water systems. Total energy saving opportunities were estimated at 12.3% of existing energy costs, with 46% of the total estimated saving opportunities related to reducing peak demands (Arora, LeChevallier and Barrer 1996). Table 5.12 lists opportunities identified to reduce peak demands.
- The Town of Fairfield Water Pollution Control Facility saved over $22,000 by participating in the ISO New England 10-minute load response program (Boman 2003).

In many of these examples, a higher level of automation has helped to support these types of energy management strategies.

Table 5.12
American Water estimated saving opportunities – 1996

Item	Installed Cost	Annual Savings	Plants
Backwash Off-Peak	$0	$10,600	Hays Mine, San Katy
Fill Washwater Tanks Off-Peak	$0	$2,200	Illinois River
Solids Dewatering Off-Peak	$0	$5,100	Hays Mine
Alternate Flights in Degritter Tank	$1,000	$1,000	Hays Mine
Operate only 1 Washwater Pump	$0	$200	Milton
Predict Tank Levels with SCADA to Manage Demands	$50,000	$27,100	Hays Mine
Minimize Peak Demand	$2,000	$5,800	Illinois River
Add Time-of Use Metering and Manage Demand	$5,000	$21,900	San Katy
Operate Diesel Pump to Minimize Demand	$0	$6,000	Milton
Operate Pump Stations Off-Peak	$2,400	$68,000	Hays Mine, Milton
TOTAL	$60,400	$147,900	

Source: Arora, LeChevallier and Barrer 1996.

Monitoring Your Energy Use

Equipment for monitoring power consumption can be incorporated into automation systems. With energy cost constituting a large portion of plant operating costs, it is useful to know how the energy is being used. By tracking overall power consumption, real-time control of demand can be implemented. Automation systems can be programmed to alarm when demand is becoming high, and to perform pre-programmed load shedding and load shifting.

By tracking the amount of water pumped per kWh, automation systems can help identify the most efficient pumps and other equipment, and schedule their use. Pump efficiency tests can be performed offline, with the controls programmed to favor the more efficient pumps.

AwwaRF report *Best Practices for Energy Management* identifies 18 energy management best practices for water utilities, ranked in order of priority, and presents several examples of each practice. Implementation of a SCADA system to track real-time operation and demand information was ranked #4, and Best Practice titled: Monitor and Analyze Load Profiles for Least-Cost Production, was ranked #2.

- A study by the California Energy Commission and EPRI that considered energy optimization at two water treatment plants and two wastewater treatment plants found that three of the plants could benefit from adding an energy management system or enhancing the current SCADA system to provide real-time demand monitoring, operator alarming, and load shedding. Estimated combined annual savings were $37,300 (EPRI 2001).
- The Erie Water Authority installed power use monitoring equipment at its Horner Pump Station, allowing the utility to determine and track the wire-to-water efficiency of each pump/motor combination. The three pumps were found to have wire-to-water efficiencies of 85.2%, 75.3%, and 84%. It was found that in the existing mode of operation the least efficient pump/motor unit was being used more than the more efficient ones. Savings were realized by shutting off the least efficient pump/motor (Porter 1996).

This research tends to indicate that there are more opportunities for effective power monitoring at water treatment plants. Again, enhanced automation for unattended operation could result in energy costs reductions as a secondary benefit.

Considering VFDs for Control

The use of VFDs in some processes can significantly improve energy efficiency. Although VFDs are electrical equipment, their use is briefly mentioned here, since they are often used for automation and control elements. Sometimes they are required to effectively control certain processes or equipment. Generally, systems that have varying flow demands and use valves for flow control or that include full-speed motors whose speed can be reduced while meeting the demand can benefit from using VFDs.

Financing Opportunities

Incorporating improvements in energy efficiency can provide opportunities for energy utilities or other entities to assist in paying for financing a portion of the project costs. Most electric utilities focus these programs on high-efficiency motors, VFDs, HVAC, and lighting improvements. However, some funding for automation may be available.

- The El Dorado Irrigation District conducted a demand response engineering analysis to identify opportunities for implementing demand response measures. It found that by shifting loads to off-peak hours and participating in the electric utility's Critical Peak Pricing and Demand Bidding Programs, it was eligible to receive over $50,000 in technical assistance for the Demand Reduction Study, over $50,000 for performance-based incentives, and $15,000 for participating in the program for a complete season, as well as savings in energy and demand charges by shifting loads. The technical and performance based incentives are tied to the reduction in peak load through operations, and can be used for software, controls, and equipment (Reely and House 2004).
- A study by the California Energy Commission and EPRI that considered energy optimization at two water treatment plants and two wastewater treatment plants identified 23 energy conservation measures at the plants, with an estimated annual savings of $564,580. Total capital costs were estimated at $621,250 with potential rebates of $547,191 available to the water/wastewater utilities from the power utilities. A portion of the savings and rebates was for load shifting and operational changes (EPRI 2001).
- The Moulton Niguel Water District received over $30,000 in cash rebates from San Diego Gas & Electric for installing variable frequency drives on its wastewater system (CEC 2003b).
- The Erie Water Authority obtained a grant from the New York State Energy Research and Development Authority for a study on how to use power monitoring data to improve efficiency. The study included installation of power monitoring equipment, measuring pump efficiencies, and development of decision support software (Porter 1996).

An estimated 4 percent of the energy used in the United States is consumed by the Water Sector (Means 2004). Energy efficiency in water and wastewater systems is the subject of considerable research. As a more thorough understanding of the subject is gained, more opportunities will be established for saving both energy and the associated costs. Major researchers include AwwaRF, The California Energy Commission, the EPRI Municipal Water & Wastewater Program, the American Council for An Energy Efficient Economy - Energy Efficiency in the Water and Wastewater Sectors, and the Department of Energy.

SUMMARY

This chapter includes descriptions of representative WTP processes and the associated unattended operational considerations. Although only representative processes were reviewed, the authors are optimistic that this will provide a framework and approach for utilities to consider additional individual processes at their treatment facilities. The goal is to stimulate ideas for considering factors important in developing the business case analysis covered in the following chapter.

CHAPTER 6
ASSESSMENT METHODOLOGIES

INTRODUCTION

How much does "advanced automation" cost? What are the benefits? Will the project pay for itself? If so, how long will it take? These are common management questions. The goal of the preceding chapters was to provide information to assist in answering these questions. Factors to be considered in planning for automation of a water treatment plant included: monitoring and staffing requirements, automation system elements, identifying automation improvements by process area, engineering and construction cost estimating, cost and benefit identification, net present value (NPV) economic cost-benefit analysis, and approaches for incorporation of intangibles using a "balanced approach."

This chapter presents a methodology for combining those elements into an assessment of the costs and benefits of automation projects and for developing an easy-to-understand automation "business case." The methodology considers both tangible and intangible costs and benefits of automation that enables unattended plant operation. This addresses situations where automation may be difficult to justify solely on a return on investment basis, but may still be desirable for other reasons. Information is provided on recommended minimum automation requirements, cost estimating, and potential benefits; the technical approach to automation must be developed considering utility-specific requirements and using engineering judgment. This report is not intended to be a design guide for automation, but to be used as a means of evaluating automation from a business perspective.

This methodology can be used in developing a business case for helping the utility determine the true value of investment in automation. A well prepared business case presents the investment goals, analyzes financial information, identifies the benefits, and evaluates the risks. The project is likely to be endorsed by the utility management if the business case incorporates the following:

- A good return on investment
- Uses an easy to understand and credible analysis approach
- Alignment with the utility's strategic goals
- Supportable propositions
- Explanation of the risks and associated mitigation measures
- Demonstrated enthusiastic support by users

METHODOLOGY OVERVIEW

Figure 6.1 provides an overview of the elements of the business case methodology. The results of the analysis of tangibles and intangibles are combined into a project "scorecard" to be used in weighing the impacts of each.

Many automation projects are implemented because of a need to support the utility's mission (or charter). For example, the requirement to provide service to all customers within the utility's geographic area, or because management recognizes that improved service to customers is a valuable asset, although one to which it is difficult to assign a monetary value.

This methodology is demonstrated with the theoretical case study titled "Roxborough Water District Feed Upgrade Project," in Appendix A.

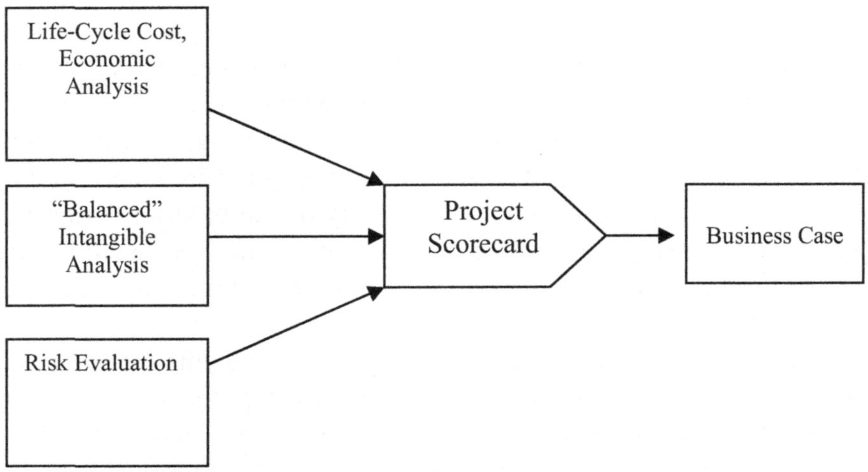

Figure 6.1 Automation business case methodology elements

METHODOLOGY STEPS

Figure 6.2 is a flow chart of the recommended steps in performing the analysis.

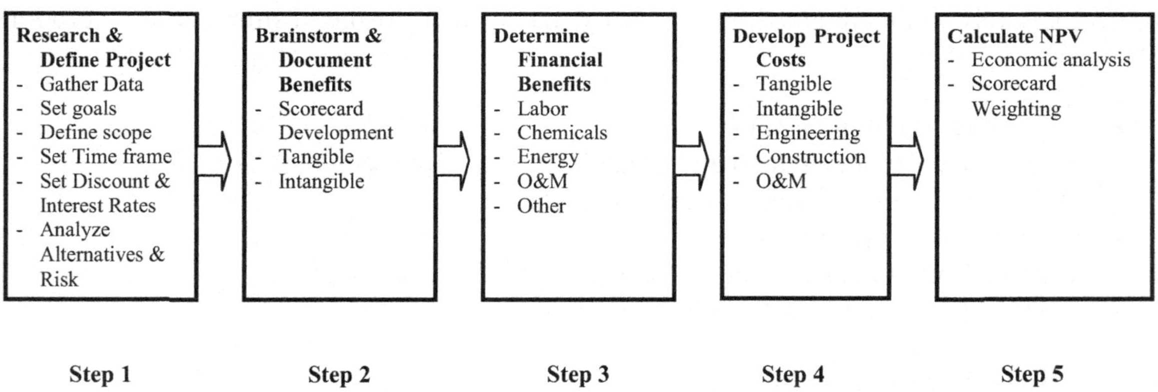

Figure 6.2 Business case analysis methodology steps

Step 1 – Research and Define the Project

A key to properly estimating the business impacts of a project is to clearly define the project elements in sufficient detail to be able to quantify costs and estimate the potential benefits.

The definition of the project may include an overview of existing or planned facilities, process characteristics, automation improvement objectives, staffing levels, budgetary constraints, and general constraints on the improvements.

If one of the goals of the economic analysis is to determine the costs and benefits of unattended operation, the economic analysis should be limited to only the items and considerations needed to incrementally move from attended to unattended operation.

For example, a high level of automation may already be provided for a process such as filtration, even though the facilities are attended around the clock. Remote monitoring by computer may be limited to the communications equipment necessary for the unattended operations. It may be logical for the analysis not to include the costs of filter automation, since these would be provided for fully attended operation.

Data Gathering. Because it is difficult to find industry benchmark information on benefits and savings relevant to the analysis, it is recommend for existing facilities, to base projected savings as a percentage of current or historical costs. Chapter 5 outlines some of the potential areas and percent savings that may be appropriate. One way to begin estimating the impacts that a project may have is to collect baseline information on the project or facility under review. Data should include the following:

- Existing staffing levels
- Prior energy bills, with usage and rate structure
- Chemical cost and use
- Production data, particularly if the planned improvements will contribute to increased capacity
- Planned improvements that add controls or rotating equipment, or increase staffing requirements
- Data on maintenance costs for instrument calibration, software maintenance, outsourced services, and upcoming expenditures associated with maintenance of automation

Establish the Project Parameters

Project-specific considerations associated with economic life-cycle analysis include the project time-line, when the costs and benefits will occur, the cost of money, and any budgetary or staffing constraints.

Project Time-Line. The project starting date, or base date, which may include the planning and detailed design, and duration of the construction and of the post-acceptance period after the improvements are implemented should be established. The base date is the date of reference for all present value calculations, and should be expressed as the year only (e.g., 2007).

An important consideration is the project time line or the life cycle duration that will be used for the analysis. As described in a previous chapter, automation projects typically have a shorter life expectancy than capital improvements such as concrete, pipelines, and major civil or mechanical projects. Computer hardware and software have a life expectancy about 3 to 5 years; and instruments, wiring, and other control devices 20 years.

These examples, the spreadsheets in Appendices A and C, and on the attached CD are based on a typical automation project life expectancy of 10 years; however, this time frame can be modified depending on the project characteristics.

Cost-Benefit Profile. An important part of the analysis is documentation of the anticipated life cycle costs and benefits. Figure 6.3 shows typical system costs as control systems age. As indicated, the costs are high when the system is first installed, and drop after startup, followed by a period when costs are predominantly for maintenance. As the system continues to age, maintenance costs rise and the original components of the system may no longer meet demands. This type of profile can be useful in estimating the timing of the costs and benefits during a project's life.

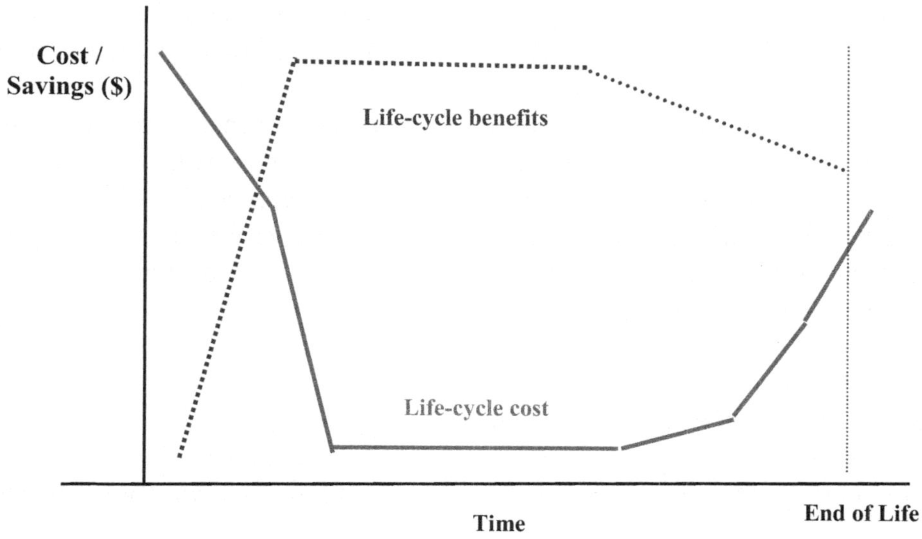

Figure 6.3 Typical profile of life cycle costs and benefits

For the purposes of the economic analysis and the spreadsheets in the attached CD, all expenditures or benefits are assumed to be realized at the end of the year being considered.

Interest, Inflation, and Discount Rates. To compute net present value, future benefits and costs must be discounted to a specific date, normally to the date when the analysis is performed. This is an important aspect of the analysis, particularly where several alternatives are considered that trade higher initial capital costs for increased benefits, or reduce long term operating costs.

In developing an appropriate discount rate, one approach is to consider the impacts of the cost of money and the inflation rate on the project. For the purposes of the economic analysis, a real discount rate will be used which can be approximated by subtracting the expected inflation rate from the nominal interest rate or the cost of money.

Figure 6.4 provides data from the Federal Government on the Consumer Price Index (CPI) which is considered a reliable indication of the general inflation rate. Federal Guidelines (Circular No. A-94) indicate that because of the uncertainty associated with predicting inflation, analysts should avoid making complicated or extensive assumptions about the general inflation rate.

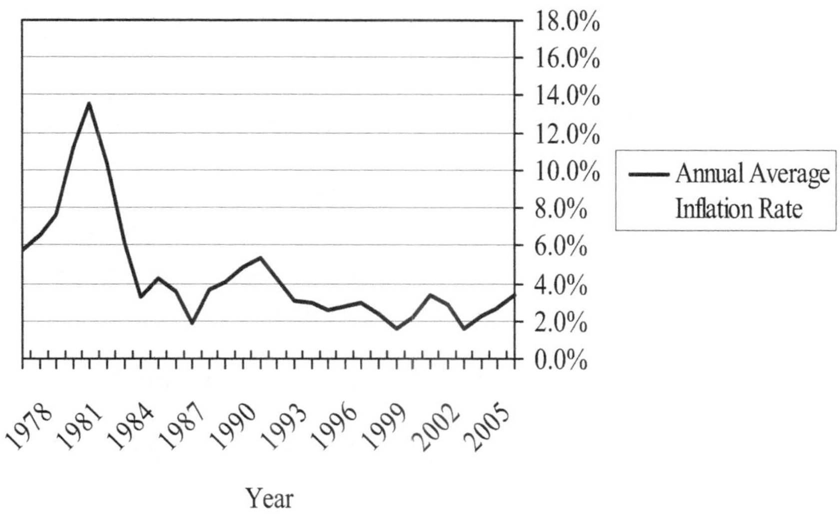

Source: US Department of Labor 2006.
Figure 6.4 Inflation rate

Figure 6.5 provides data from the Federal Reserve Board on the Federal Funds rate, which provides an indication of the interest rates that the Federal Government charges banks. Typically, long-term financing rates are the Federal Funds Rate or Prime Rate plus a small percentage to make the financing attractive to investors. Actual interest rates that municipalities pay for the cost of money may be higher or lower depending on the funding sources and credit worthiness. The Federal Funds Rate does provide a relative indication of interest rates and their trends.

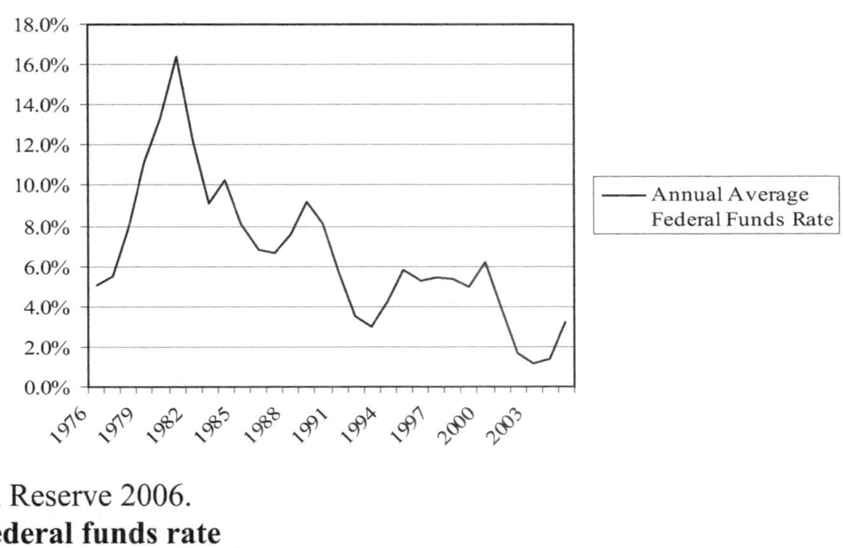

Source: Federal Reserve 2006.
Figure 6.5 – Federal funds rate

Each utility should consider its specific circumstances regarding the cost of money and the anticipated inflation rate in order to establish a reasonable discount rate for the analysis. This report assumes a 20-year average of the CPI (3.0%) and Federal Funds Rate (5.1%).

Other Considerations. Identification of processes and project constraints is important. This could include the remoteness of the facility, accessibility, communications with off-site operators, availability of qualified support, system construction and cut-over constraints, technical skill levels of staff, additional spare parts and servicing for maintenance of new automation equipment, and limitations imposed by regulations. Are there specific things that must be documented to satisfy local regulators?

Budgetary constraints or personnel qualifications which may have an impact on the alternatives or the ongoing training or maintenance costs should be identified. In many cases, capital budgets may not be available for the automation improvements even if the long-term life-cycle costs would result in a favorable project economic analysis.

Define Project Goals, Strategies, Objectives, Approach, and Scope

During this step, the project is defined and agreement is developed among stakeholders on the business goals, the strategies for achieving the goals, and the resulting project scope. The business goals will describe the issues being addressed and the expected benefits to the utility. These are the primary reasons for proposing the project. The strategies describe the approach being taken to realize the benefits. The project scope defines the work to be done and the limits of the project. A clear scope is an essential pre-requisite for the cost analysis conducted in Step 4.

At this step it is useful to create a conceptual diagram of the control system improvements planned. Another useful tool is to prepare a simplified process schematic or P&ID identifying the major elements and scope of the work as well as descriptions of the control system improvements and operational strategies. Several of the case studies in Appendix B provide representative examples of simplified process diagrams and control descriptions.

Reference Figure 3.1 and the cost database in Appendix C for an overview of the elements that may be encountered in a typical treatment plant automation project, regardless of the plant processes. For example, an auto-dialer could be installed to notify plant operators. This would be a common cost item, with no differences depending on plant size.

Some automation improvements require not only elements directly related to automation, but additional process, mechanical, or electrical improvements, such as electric valve actuators on filter valves or variable frequency drives on pumps. The need for such items should be determined as part of the analysis.

Chapter 5 provides information on typical automation requirements for unattended operation and provides ideas for identifying the potential cost savings, risks, and mitigation measures.

Evaluation of Alternatives

The analysis should consider alternative methods of achieving the project goals. For example, alternatives for upgrading filter controls may include the following: (1) do nothing, (2) purchase new equipment (3) upgrade or renovate existing control equipment, or (4) maintain functionality by increased maintenance.

When selecting alternatives for economic evaluation, consideration should be given to reliability or redundancy and opportunities for savings in energy, chemical, or labor. Important considerations might include alternatives that trade recurring savings for higher or lower initial capital expenditures.

Risk Analysis

It is recommended that the automation systems incorporate elements that minimize the risks associated with equipment failures, and that any associated additional costs be included in the analysis. Typical considerations may include redundant instrumentation in areas where failure of an instrument would have a significant impact on the system's operation or water quality.

Other risks to be considered are those associated with the cut-over methods and the use of new, unproven technology. Chapter 4 identifies various risk considerations associated with unattended operation and provides recommendations on factors to be considered in planning for unattended operation. Chapter 5 presents information on risk considerations and potential mitigation measures for several of the more common water treatment plant processes.

Step 2 – Brainstorm and Document Benefits

Some areas where project benefits may be identified are presented in Table 6.1. The primary benefits will be obvious since these are the project drivers. Often the secondary benefits can be achieved at little additional cost simply by designing the project to support other utility goals, such as security, improved emergency response, and reduced down time.

Table 6.1
Example areas for discovering benefits

Item	Benefit description	Target benefit relative importance	Strategy
Financial			
A1	Reduced design and startup costs through applying new standards.	10%	Applying automation standards, and tested modular software reduces design and startup costs. Does this project also establish automation standards that will be beneficial for later downstream projects?
A2	Reduced energy costs	5%	Apply "time of day" pumping to match times of lower energy costs. Use VFDs to improve pumping efficiency. Are there energy strategies that have been incorporated into the advanced automation design that will generate energy savings?

(Continued)

Table 6.1 (Continued)

Item	Benefit description	Target benefit relative importance	Strategy
Customer			
B1	Improved security	5%	Implement increased physical security monitoring (gates, fences, doors) and computer security (fire-walls, encryption, bio-security). Reduced risk of terrorist/hacker intrusion compromising water to customers. Is security addressed on the project that will contribute to risk mitigation? Does this align with a wider utility strategic plan?
B2	Improved utility image	2%	Customers perceive utility as proactive in protecting critical water infrastructure. Adding security features (as an example) to a project allows the utility to be seen by the customer as caring about public health, and pro-active in protecting consumers.
B3	Improved emergency response time.	2%	Use improved process monitoring to flag and respond to abnormal conditions early. Protecting consumer health is a utility strategic goal.
B4	More consistent water quality	5%	Advanced automation improves the quality and consistency of the water delivered, particularly under abnormal conditions.
Business Processes			
C1	Reduced Operator labor costs through unattended operation.	45%	Change from central facility monitoring for three shifts, to "On Call" for two of three shifts. Savings accrue because of reduced Operator costs. This is a major driver for the project. Calculate the difference between "On Call" operator costs and a "fully staffed" control room.
C2	Reduced operation costs by automating routine Operator procedures.	5%	Automate routine procedures to free Operators for other duties. Examples include: Flow pace chemicals, Provide feedback control of chlorine residual.
C3	Reduced down time/emergency call-out costs through improved monitoring and diagnostics	5%	Improved monitoring of equipment and equipment diagnostics provides warning of failures before they occur. Costs are reduced because of reduced emergency callouts.
C4	Reduced maintenance costs through improved monitoring and diagnostics.	10%	Improved monitoring of equipment and equipment diagnostics provides warning of failures before they occur. Costs are reduced because maintenance is planned rather than developing into an emergency event. Better on-line monitoring can include: Pump efficiency calculation and alarming, Valve torque/travel, end of travel monitoring and alarming, Well efficiency calculation and monitoring, Communication efficiency and alarming. Estimate the labor savings per year associated with providing scheduled maintenance

(Continued)

Table 6.1 (Continued)

Item	Benefit description	Target benefit relative importance	Strategy
C5	Reduced maintenance costs by installing new automated equipment.	5%	New equipment installed as part of the project replaces older less reliable equipment. Cost savings through reduced maintenance costs.
C6	Reduced Operator error	2%	Provide the Operator with accurate information on abnormal conditions to ensure a correct diagnosis is made immediately. Cost savings occur if only one response call is needed by utility staff. Savings if conditions indicate response can be delayed to the next attended shift.
Learning and Growth			
D1	Staff training improves decision making	2%	Training provided by the project increases skills of utility staff and their ability to respond to abnormal conditions.
D2	Improved staff morale	1%	Staff training improves morale. Staff perceive utility investing in improving skills.
D3	Improved Utility knowledge base	2%	Project documentation progresses the utility strategic goal of documenting Operator knowledge base. Assists training new operations and maintenance staff by adding to the electronic O&M.

This step involves discovery and documentation of benefits. Different stakeholders will have different perspectives on the project and each will valuable input to the discovery process. The stakeholders for an advanced automation project could include operations, water quality control, maintenance, engineering, systems and management personnel.

Part 1 Brainstorm with stakeholders on all possible project benefits, both tangible and intangible, starting with the primary drivers that created the project. This is commonly done in a workshop environment using Table 6.1.

Part 2 Select five to ten key benefits to be used in developing the business case and present an objective evaluation of the project to management. The stakeholder group should rate each benefit to reflect its expected contribution to the total project benefit. A simple example follows in Table 6.2.

Table 6.2 Sample benefit ratings

Project benefit	% Contribution
Reduction in operator labor hours	45
Reduction in energy costs	5
Reduced maintenance/ improved reliability	20
Improved water quality	20
Improved security	10
Total benefit	100 %

Note that the first three benefits are "tangible." That is, a dollar savings can be attributed to them. The last two are "intangible" benefits since it is difficult to assign a dollar value to them. The percent value assigned to each benefit is not a calculated value but rather a stakeholder consensus of the importance the group puts against the benefit. This approach properly recognizes the importance of both intangible and tangible benefits.

The workshop approach to driving out benefits has the added advantage of documenting stakeholder expectations (and hence requirements) and getting buy-in to the project goals. Some organizations may prefer a different approach, such as one-on-one interviews to collect input.

Some organizations may not want to heavily emphasize the intangible benefits or assign them a significant weight. In such cases, the general framework is still applicable however, it may be worth considering the intangibles in the overall analysis but to assign them a weight of zero in financial considerations.

Step 3 - Analyze Financial Benefits

This step assigns a dollar value against each tangible benefit. The following are areas where tangible benefits may be realized. The majority of these benefits would occur after the commissioning and acceptance of the project:

Labor Cost

This is an area of potentially significant benefits or cost savings if the automation enables unattended operation. One approach is to look at the net reduction in staffing levels or a reduction in the number of shifts, which would lead to a reduction in annual labor costs.

Chemical Cost

Automation can result in better control of chemical dosages and thus in a net reduction in chemical costs. The baseline data on chemical consumption costs should be reviewed considering the potential percentage reduction in terms of annual costs.

Energy Cost

If the automation improvements in conjunction with the improvement needed for unattended operations result in improved plant operation, there may be opportunities for energy savings in areas such reduction in pumping as a result of reduced backwash volume or waste recycling. Shifting the pumping to off-peak hours or sequencing the operation of large equipment may minimize demand charges.

Miscellaneous Costs

This category can include cost items such as elimination of recurring computer hardware and software maintenance costs or any future hardware costs. Other miscellaneous costs including savings in process water, reduced spare parts inventories, etc., should also be considered.

Revenue Enhancement

Any potential improvements in plant capacity or the ability to maximize the plant output could result in revenue from increased water delivery. There are a number of methods for deriving these values, including the following:

- Using published cost savings data for existing similar facilities. For example, AwwaRF report "An Evaluation of Streaming Current Detectors" showed that adding a streaming current detector for filter coagulant control could save an average of 11% in chemical costs during periods of stable raw water turbidity for the eight plants studied. Using these types of figures and the plant's past chemical costs a rough estimate can be made of the potential.
- Analysis of past utility costs and a best estimate of savings. This less rigorous method can lead to realistic estimates of saving by estimating the percentage of historical costs that would be changed by the project. For example, this project may reduce maintenance costs by 15 to 20%. Using the utility's past annual maintenance costs and the 15% or 20% reduction, the anticipated savings can be calculated.

In both approaches described above, it is important to document how the costs and percentages were derived. Where adequate data are not available, an alternative approach is to use data and metrics provided by utility personnel. The task can be completed in a workshop environment as described in Step 2, above. Each tangible benefit is examined and the costs and savings are calculated based on the labor, materials, and energy costs. If staff members have been with the utility for at least two or three years, a wealth of collective information is available. A conservative estimate of savings should be used. It may be appropriate to provide a range of potential savings in the analysis. The basis and data for each calculation should be recorded in the same manner as for the other approaches.

The success of this method relies on having the appropriate experienced staff together at the workshop. Examples of savings calculations are provided with the example project in Appendix A.

Step 4 – Develop Project Costs

Cost Estimating

There are a number of costs associated with planning, designing, procurement, installation, operation, and maintenance of automation systems. One challenge is to determine what costs should be considered in the life-cycle cost analysis of unattended plant operation. Some costs that are necessary only to provide a base level of control may or may not be included in the analysis. To identify the real costs and benefits of transitioning to unattended operation, only the essential elements should be considered.

In many cases, it may be true that the incremental costs associated with a change in operational strategy may have a small incremental cost over the automation provided for a fully attended but highly automated plant.

Only those costs that have an impact on the decision and represent a significant expenditure or benefit are needed to make valid business case decisions. Any sunk costs, ones that have previously been incurred that can't be avoided by a future decision, should be excluded from the analysis. For example, the cost of pneumatically controlled filter valve actuators that are being replaced with electric actuators, regardless of the project being considered, is considered a sunk cost.

Tangible Costs

The tangible costs associated with an automation project may be classified as acquisition costs which include planning, engineering, and construction costs, and operating and future costs.

Planning and engineering. This includes life-cycle costing, business case development, master planning, pre-design, detailed design, procurement services, and engineering support during construction.

Construction. Includes costs associated with the procurement, installation, and commissioning of the automation systems. A detailed approach to estimating construction costs which may include computers, control panels, instrumentation, control devices, software integration services, testing, training, interconnecting wiring, physical improvements necessary to support the automation like valves, chemical feed equipment, actuators, etc., is presented in Chapter 3.

Operating and Future Recurring Costs. These costs include ongoing expenditures after the project has been commissioned, such as instrumentation calibration and maintenance materials, repairs, spare parts, software support, software upgrades, computer system maintenance, and ongoing training, which are often not included or are underestimated. Utility costs may be worth considering if there is an increase in energy or water costs.

It is useful to categorize these costs into single, or one-time, costs and recurring costs. To simplify the net present value calculations, it is recommended that the recurring costs be considered to occur at the end of each year.

Chapter 3 also includes a detailed methodology and model for estimating the annual maintenance costs for automation.

Intangible Costs

Intangible costs that may result from an automation project are difficult to quantify, and could include resistance to change, the need for organizational changes to support new technology, and other considerations.

Step 5 – Calculate Project NPV

In order to perform a NPV life-cycle cost analysis, all project related costs and benefits need to be identified according to when they occur. It is an accepted practice to simplify the analysis to treat all the costs or benefits as occurring at the same time of the year. FEMP rules (10CFR436) allow for single and recurring costs to be discounted either from the time of occurrence or at the end of the year in which they occur. In order to simplify the analysis, it is recommended that the costs and benefits be considered to occur at the end of the year in which they occur.

Computer programs or spreadsheets, like Excel (Microsoft, WA), typically specify when the expense occurs. The NPV function in Excel (Microsoft, WA), considers recurring costs to occur at the end of each year.

Economic Analysis

Several spreadsheets to support the economic analysis are in Appendix A and on the CD. The sheets are organized by costs and benefits and in the electronic spreadsheet files are linked to a summary sheet that calculates the lifecycle costs-benefit for the project. The summary identifies the total costs by category per year and then provides a graphical depiction of these costs and benefits.

Scorecard Weighting

In the event that the NPV of the project is less than desired, it may be appropriate to consider the relative weightings developed in Step 2 as a way to provide further insight into the value of the project. One question could be "why perform Step 2 before the NPV and why not wait to see if the NPV is desirable before looking at the intangible benefits?" It is recommended that Step 2 be conducted prior to calculating the NPV since it will provide a more objective ranking of the intangible benefits.

The weighting can be used to adjust the financial numbers by increasing the benefits or normalizing the costs. One approach is shown in the example in Appendix A.

DEVELOP THE BUSINESS CASE DOCUMENT

A business case should be prepared to obtain executive buy-in and commitment for the project. It is likely to be compelling if it concisely answers the following key questions:

- What is the project intent and scope?
- Is the project aligned with the utility's strategic plan?
- Does it meet one or more of the strategic plan's objectives?
- What will the project cost?
- What are the prospective benefits?
- Do the benefits justify the cost?
- Are the benefit assumptions reasonable?
- What are the consequences of not proceeding with the project?
- Is the risk acceptable?
- Do stakeholders enthusiastically support the project?

In some cases the calculated NPV for the tangible benefits may not result in a satisfactorily high net positive return. This does not mean that the project is not a worthwhile one. If there are important intangible benefits (for example improved security, or more consistent water quality) management may accept the lower return because of the strong value of the indirect benefits.

As part of Step 2, stakeholders attributed a quantitative value to the importance of the tangible benefits. If intangible benefits are accepted, the NPV can be judged against a percentage of the total cost, where the percentage is the percentage of importance of the tangible benefits.

Business Case Outline

A business case should be built around five or six strong benefits, tangible or intangible. Adding a long list of further, weaker benefits tends to dilute the effectiveness of the argument. Be concise. Avoid making the readers struggle through pages of text to find the nuggets they need to make a decision.

The following is an outline of an example business case document:

1. Cover
2. Table of contents
3. Executive Summary
   - Intent of project
   - Recommendation
   - Benefits
   - Return on investment
   - Risks
   - Next action
4. Project Analysis
   - Methodology for gathering input, and identify contributing staff members
   - Project scope and assumptions
   - Top benefits
   - Benefits related to the Utility's strategic plan and objectives
   - ROI analysis summary. Reference the cost and benefits worksheets in the Appendices
5. Risk analysis
6. Conclusion and next actions
7. Appendices
   - Cost and benefit worksheets
   - Project support information

SUMMARY AND RECOMMENDATIONS

The goal of this report is to provide an approach to evaluating the costs, benefits and risks associated with unattended WTP automation. A major conclusion drawn from the study was for water utilities to employ recognized industry methodologies for justifying automation projects. The literature search indicated that a formal approach has been conspicuously lacking in the water industry in the past. Developing a credible business case helps clarify project goals and scope and allows management to make informed decisions.

One of the goals of the methodology presented in this chapter is to provide a framework for approaching the analysis and other chapters provide background information to support the analysis. The decision to operate unattended is a complex one and developing a sound business case can be challenging.

For water utilities that are considering the costs and benefits of unattended plant operation the following recommendations should be considered during the evaluation:

- Investigate state regulations and identify any regulatory constraints on unattended operation.
- Carefully define the scope and goals of the automation project.
- Evaluate the risks and consequences associated with the potential failures of automation.
- Provide a level of margin between the operational and process goals and the regulatory limits on plant operation.
- Develop a cost model including the capital and operating costs of the automation. Do not underestimate the construction costs and the ongoing operations and maintenance costs.
- Define the benefits of automation both tangible and intangible through brain-storming sessions with operation and maintenance staff. Quantify the tangible benefits and rate the importance of the intangible benefits. Use conservative estimates of expected savings.
- Build consensus and management involvement early in the development of a business case for automation improvements.
- Develop a project business case that can be presented to management. Include both a benefit and a risk analysis. Recognize that automation improvements may be difficult to justify based solely on tangible benefits.
- Design an automation system to support unattended operation.
- Employ industry best practices for engineering, contracting for services, and procurement. As described in Chapter 4.
- Establish a program to better collect historical data on plant production, energy utilization, chemical costs, and labor costs prior to completing the economic analysis.

FUTURE RESEARCH

The decision to operate water treatment plants in an unattended manner is a complex one involving more issues than economics alone. The research team encountered many cases where the financial benefits were not the deciding factors in the decision whether to operate unattended. In some cases where there was a desire to perform an economic analysis, the data was not available to support a thorough evaluation. In another case, although the utility had adequate automation to support unattended operation, due to regulations they did not operate in that mode. To address some of these overarching concerns, the following topics are suggested for future research:

- Develop information or methods for better communication to financial decision makers and regulators that complete automation can be a good thing. This may come in the form of a communications project.
- To assist water utilities in performing an economic analysis of their situation, it would be useful to develop a framework for economic and performance data collection. The goal would be to develop approaches that utilities can take to structure data gathering, historical data storage and performance metrics so that performance evaluation can be done on an ongoing basis. This information would allow utilities to better assess potential savings from complete automation.

APPENDIX A
EXAMPLE BUSINESS CASE ANALYSIS

EXAMPLE NPV ANALYSIS

ROXBOROUGH WATER DISTRICT COAGULANT FEED UPGRADE

The following example shows the use of the Business Case Worksheets provided on the attached CD to arrive at a credible basis for writing the business case. It is written for a **theoretical** upgrade project at the Roxborough Water District (RWD). Roxborough intends to automate the coagulant feed to the filters with two main objectives in mind. The first is to reduce coagulant costs, and the second to provide a more consistent water quality. The raw water feed through the plant can vary daily depending on demand, the number of filters in operation, and the raw water quality. RWD management requires new capital projects to have a business case developed that justifies the District's investment. Both tangible and intangible benefits are considered.

Currently the operators manually set the feed rate slightly higher than needed to allow for variations in flow and raw water quality. The operators monitor the post filter turbidity during the day shift (via SCADA) to ensure the treatment process is performing adequately. During the night the operators rely on being paged by the SCADA system if a turbidity alarm occurs. Generally the coagulant feed rate is adjusted twice during the day to compensate for variations in the raw water. An adjustment is also made after a change in the number of on-line filters. Adjustment requires walking out to the chemical building and adjusting the stroke of the coagulant pump. This pump is fitted with automatic stroke control, but this feature is not currently being used. The plant is operated 24 hrs/day and 7 days/week with at least one operator on-site at any given time.

The project proposes adding a flow meter and streaming current detector upstream of the coagulation point. These measurements will provide feed forward signals to a coagulant feed rate controller which will be programmed into the existing filter PLC. This will send a dosage signal (4 to 20mA) to the feed pump stroke controller.

The following approach to developing the business case was followed.

Step No. 1 - Research and Define the Project

Step 1A - Project Research and Data Gathering

Staff researched and gathered data for the following:

- Historical chemical costs: Operations staff collected data on the coagulant chemical costs and quantities over the previous three years based on billings from the chemical supplier. The calculated average historical coagulant cost was $25 /mgd.
- Chemical savings estimate: Engineering researched documented cases where automatic control of coagulant had been implemented for similar agencies. The literature indicated a possible chemical savings of approximately 11 percent. The savings were for utilities with variable flow rate and raw water quality, controlling coagulant feed based on feed forward of raw water flow and streaming

current. A conservative savings rate of 10% was chosen for the economic analysis.

- Maintenance costs: Discussion with the maintenance department resulted in an estimate of 2 hours per month for calibration and cleaning of the instruments.
- Project life: The project life was considered to be approximately 10 years. This was based on the expected life of the streaming current detector, which was the most expensive component.
- Cost of money: The District cost of money for capital projects is 7.1%.
- Estimated Inflation Rate: 3.0 %
- Calculated Effective Discount Rate: 4.1 %

Step 1B - Define Project Goals, Strategies, Objectives, Approach, and Scope

Staff meetings were held to discuss and agree on the overall purpose and scope of the project. Operations, Water Quality, and Engineering staff agreed on the following:

Goals	Goal 1: To reduce coagulant chemical costs Goal 2: To improve the consistency of water quality, particularly during upsets in raw water quality.
Strategies	• Add only the coagulant quantity needed to meet water quality targets or goals. Control coagulant feed to automatically adjust for changes in raw water flow and quality. The WQ target should provide some margin of error between the targets or goals and the standards. Provide the flexibility to respond manually well before the standards are not met. Loss of a coagulant feed is a violation of the IESWTR requirements therefore an acute MCL violation.
Objective	• Reduce coagulant feed costs by at least 10%. • Reduce labor costs associated with operating and maintaining the coagulant feed system by 40% or more compared with historical labor costs.
Approach	The chemical feed will be automatically controlled based on measurements of the two major variables: raw water flow and streaming current. This will require installing a new flow meter and streaming current detector. These two signals will be inputs to a new electronic coagulant feed controller.

Scope	Project work needed: • Add new instruments to measure raw water flow and streaming current. • Add a program to the existing filter PLC to provide a control loop that regulates coagulant flow based on feed forward signals of raw water flow and feedback on streaming current. • Cable instrument signals to the PLC and provide an analog output signal to control the chemical pump stroke.

The description and process schematics were entered on the project Worksheets 1A and 1B in the workbook in Appendix A.

Step 1 C – Alternatives Analysis

One alternative considered was to do nothing. Another alternative considered was coagulation optimization using a neural network based optimizer. This technology was considered to be in the early stages of development and would require further development for this application.

Step 1 D - Risk Considerations

Due to the potential water quality issues that could result from a malfunction or failure of the coagulation control it was recommended that a failsafe interlock be provided in the event of equipment failure to shut down the plant. The anticipated response time for a shut down condition is 25 minutes. This duration of plant shutdown was considered to be manageable.

Step 2 – Brainstorm and Document Benefits

A meeting was held to brainstorm benefits of the project and was documented using Worksheet 2A and 2B. Five benefits were identified. Two benefits had dollar cost savings (tangible) and three had non-financial (intangible) benefits. The importance to the District of all benefits was discussed and weighted. This produced the following results:

Step 2A - Brainstorm Major Benefits Provided By The Project

	Reason	Expected benefit
Benefit 1	Chemical savings by closer control of chemical feed	Reduction in cost of coagulant per MG treated
Benefit 2	Reduced operator attention because of automation of the coagulant addition	Gained operator hours can be used for light maintenance work. Reduced nighttime callouts.

	Reason	Expected benefit
Reason 3	Improved water quality because of better particle reduction	Improved customer satisfaction
Reason 4	Early warning of coagulant equipment and process problems because turbidity alarm could be set closer to setpoint.	Reduced customer complaints when system problems occur or sudden changes in raw water quality cause upsets. Reduced risk of compliance problems.
Reason 5	Improved Operator effectiveness	Operators can focus on more important District issues while maintaining improved water quality.

Step 2B - Classify and Weight Expected Benefits

	Benefit Description	Classification	Percent of total expected benefit
Benefit 1	Chemical savings	Tangible	60%
Benefit 2	Reduce operator labor	Tangible	20%
Benefit 3	Improved consistency of water quality	Intangible	10%
Benefit 4	Early warning of problems	Intangible	8%
Benefit 5	Improved effectiveness	Intangible	2%
			100%

Step 3 – Analyze Financial Benefits

This step is divided into three parts:

Part A – Calculate savings from tangible benefits
Part B - Calculate O&M cost reductions
Part C - Document data sources

The expected savings and change in costs compared with the historical costs (collected in Step 1) were noted on Worksheet 3 and the annual savings calculated. The formula used for the calculation was documented for clarity and later review. The source of the data was noted in Part C to add credibility to the calculations.

Step 4 – Develop Project Costs

Engineering developed a capital cost estimate for the project and entered the value ($3,933) into the "NPV Analysis Spreadsheet". This included design, bidding, construction, startup and training. The capital costs were all entered into the first year of the expected 10 year project life. Quotes for the flow meter and streaming current detector were obtained from vendors. Note that the instrument maintenance costs have been entered in the "Post Acceptance Support O&M Costs" sheet for each of the 10 years.

For more complex projects refer to the construction cost estimating guidelines provided in Chapter 3 and the automation construction cost estimating model.

STEP 5 – Calculate Project NPV

To enable the NPV to be calculated, the benefits and O&M costs were entered into the "NPV Analysis Spreadsheet." Note that the instrument maintenance costs have been entered in the "Post Acceptance Support O&M Costs" sheet for each of the 10 years. The cost of money value was entered at the top of the Summary sheet.

The NPV totaled $105,477. This indicates that this project is relatively profitable over its 10 year life. In fact the project will pay back the original capital investment of $63,933 in a little more than three years. For situations where the project benefits may not offset the expenses, one approach is to account for the contribution of the intangibles by discounting the costs or increasing the estimated benefits in-line with the percentages outlined in Step 2.

The following pages provide printouts of the Excel (Microsoft, WA) workbook spreadsheets. These are provided on the CD for reference.

ROXBOROUGH COAGULATION CONTROL UPGRADE PROJECT
BUSINESS CASE ANALYSIS

Utility Name:	Roxborough Water District
Name of Project	Coagulation Control Upgrade Project
Project Location:	Roxborough Plant No.1
Evaluation Date:	January 2006
Project Service Date:	July 2006
Study Participants:	Tex Harrison, Maddona Rock, Tom Banks, William Howard
Study Date:	Feburary 2006

Utility Overview: Roxborough Water Utility treats raw water from the Domingo Reservoir and supplies 35,000 households and some light industry in the town of Roxborough and its surrounding area. The 8 mgd treatment plant is conventional consisting of raw water pumping, coagulation, sedimentation, filtration, and distribution pumping.

Project Description and Goals

(Provide a summary of the project including capacity and size. Document the goals, strategies,and objectives, Approach, and scope)

This project upgrades the coagulation chemical feed by automating the feed based on the measured raw water flow and streaming current.The projected is expected to reduce both labor and chemical costs and maintain a more consistent treated water quality.

Goals	Goal 1: To reduce coagulant chemical costs Goal 2: To improve the consistency of water quality, particularly during upsets in raw water quality.
Strategies	• Add only the coagulant quantity needed to meet water quality standards. Control coagulant feed such that it automatically adjusts for changes in raw water flow and quality. • Automate the process to reduce labor costs.
Objective	• Reduce coagulant feed costs by at least 10%. • Reduce labor costs by 40% or more compared with historical labor costs.
Approach	The chemical feed will be automatically controlled based on measurements of the two major variables: raw water flow and streaming current. This will require installing a new ultrasonic flow meter and streaming current detector.
Scope	Project work needed: • Add new instruments to measure raw water flow and SC. • Add a program to the existing filter PLC to provide a ratio controller that regulates coagulant flow based on feed forward signals of raw water flow and streaming current. Cable instrument signals to the PLC and provide an analog output signal to control the chemical pump stroke.

STEP 1 - RESEARCH AND DEFINE PROJECT

ROXBOROUGH COAGULATION CONTROL UPGRADE PROJECT
PROJECT SCOPE & PROJECT RESEARCH

System Operation Description:

(Provide a description of the plant operating strategy or control strategy descriptions.)

The proposed coagulation control upgrade project at Plant No.1 will provide controls to automatically adjust the coagulant chemical feed to the raw water based on the plant flow and feedback control on the raw water streaming current detector. This will eliminate the need for the operators to manually adjust the feed to compensate for flow changes and fluctuations in the turbidity of the water from the reservoir.

Automation Features Description:

(Provide a description of the automation features and equipment utilized to enable unattended operations.)

The coagulant feed will be paced based on the measured flow and the streaming current of the raw water, measured upstream of the coagulant injection point. A new ultrasonic flow meter and a streaming current detector will be purchased and installed, with the measurements taken to the filter PLC. The control algorithms will be programmed into the existing filter PLC.

Research:

1. Plant Staffing:

(Provide a description of the staffing level of the plant, number of shifts, number of operators)

The plant is manned 24/7.

It is estimated that the man-hours required to supervise the coagulant system, and conduct lab tests will be reduced to 0.5 hours/ day, averaged over summer/Winter. At the operator hourly rate of $75/hr:

Staffing cost reduction/day = 0.5 x $75 = $37.50

2. Coagulant savings through automation

Research of the literature (reference studies by Dentel, Kinerery, AWWA 1988) indicates that other utilities with similar variable flow and raw water quality have achieved coagulant savings between 3 and 63 percent, with an average of 23 percent. Hence a conservative savings of 11 percent was assumed

3. Maintenance Costs

Two additional instruments will be added by the project. The Districts maintenance staff were asked to estimate their monthly maintenance costs to keep the instruments operational and calibrated. Staff estimated 2 hours per month, based on a calibration check every 2 weeks, and recalibration every 3 months.

ROXBOROUGH COAGULATION CONTROL UPGRADE PROJECT
PROJECT DEFINITION - SCHEMATIC DIAGRAMS

System Schematic: (Provide a simplified schematic of the treatment plant control system and process. Attach P&ID drawings if available.)

Current Plant Schematic

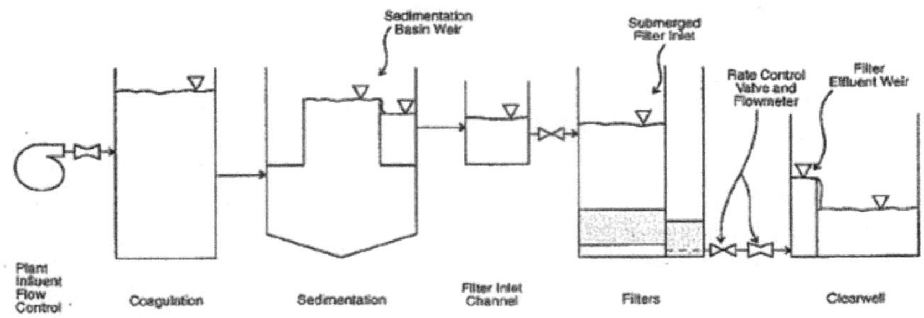

Upgraded coagulant control.

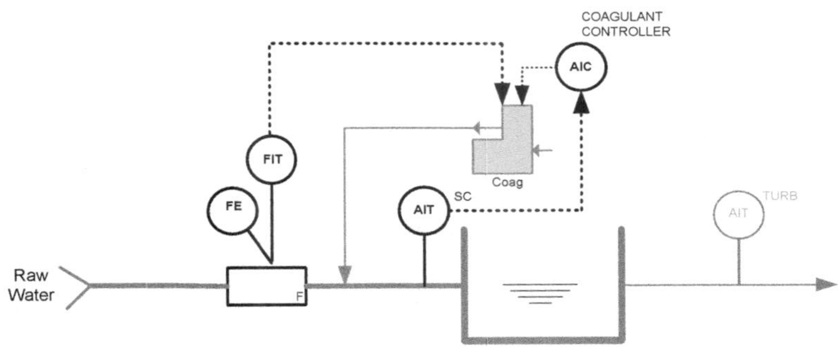

New equipment and instruments are shown bold.
New instruments include:

1. Ultrasonic flow meter
2. Streaming current detector

Note that the Coagulant feed controller will be programmed in the existing PLC.
Sufficient PLC I/O are available without the addition of further I/O cards

Figure A.1 Example Process and Instrumentation Diagrams

STEP 2 - BRAINSTORM AND DOCUMENT BENEFITS

ROXBOROGH COAGULATION CONTROL UPGRADE PROJECT

BENEFITS CHECKLIST

POTENTIAL BENEFITS CHECKLIST

This sheet is intended to suggest additional benefits of the project, that may not have originally been targeted. The following is a check list of common benefits, both tangible and intangible

1.0 Cost Savings	
1.1 Use of Automation to save operator effort.	
1.1.1	Additional monitoring reducing operator visits to process
1.1.2	Automating a routine process strategy.
1.1.3	Reduced labor through centralized control
1.1.3	Reducing operator costs by enabling "on-call" operation through off-site monitoring of process.
1.1.4	Enabling "on-call" operation by providing off-site monitoring alarming.
1.2 Reduced chemical costs	
1.2.1	Use of flow pacing chemical addition
1.2.2	Feed back control using an analyzer
1.2.3	Analyzer addition to improve monitoring
1.3 Reduced energy costs	
1.3.1	Time of day pumping
1.3.2	Use of VFDs instead of valve throttling
1.3.3	Use of more efficient motors
1.3.4	Ability to change between tarifs or energy sources.

1.4 Reduced construction costs	
1.4.1	Use of design/build
1.5 Use of methodologies which save costs for follow-on projects	
1.5.1	Development of standard design that can be re-used
1.5.2	Development of PLC standards/programs that are reusable
2.2 Reduced Emergency call-out effort	
2.2.1	Providing off-site Operators with improved process information/ more detailed alarm information so better emergency decisions can be made. Reduced call-outs.
2.3 Reduced maintenance costs	
2.3.1	Provision of condition monitoring to anticipate equipment problems.
2.3.2	Installing more reliable equipment
2.4 Reduced down-time by improving reliability.	
2.4.1	Adding redundancy to equipment
2.4.2	Adding redundancy to control system
2.4.3	Adding redundancy to HMI
2.4.4	Installing more reliable devices.
2.4.5	Adding strategies for failure detection and automatic fail-over to backup systems

STEP 2 - CLASSIFY AND WEIGHT EXPECTED BENEFITS

ROXBOROUGH COAGULATION CONTROL UPGRADE PROJECT

STEP 3 - BENEFITS

Document the major reasons/ benefits provided by the proposed project.

	Reason	Expected benefit	Classification	Percent of Total
Benefit 1	Chemical savings by closer control of chemical feed	Reduction in cost of coagulant per MG treated	Tangible	60%
Benefit 2	Reduced operator attention because of automation of the coagulant addition	Gained operator hours can be used for light maintenance work.	Tangible	20%
Reason 3	Improved water quality because of better particle reduction	Improved user satisfaction	Intangible	10%
Reason 4	Early warning of coagulant equipment and process problems because turbidity alarm will be set closer to setpoint.	Reduced user complaints when system problems occur or sudden changes in raw water quality cause upsets. Reduced risk of compliance problems.	Intangible	8%
Reason 5	Improved Operator effectiveness	Operators can focus on more important District issues while maintaining improved water quality.	Intangible	2%
				100%

STEP No.3 - O&M COST AND BENEFIT CALCULATIONS

This sheet has two calculation tables, "Savings" (Parts A) and "O&M Costs" (Part B). The third table (Part C) documents the source of the data used as the basis for the calculations.

Part A - Savings Calculation for Tangible Benefits

Notes:

Key Metric:	Measurement which is the basis of the calculation		Change:	The difference in value between colums A and B
Base Period:	Value prior to implementation of project		Annual $savings formula:	How the savings are calculated
New:	Estimated value after project implemented		Annual savings:	The results of the calculation in Column E

	Description	Key Metric	Base period A	New B	Change C	Units D	Annual $Savings Formula E	Annual $Savings F
Benefit 1	Reduced coagulant use through better control	Coagulant cost/ MG	$25	$23	$3	MG	365days x 8 MGD x "C"	$7,300
Benefit 2	Reduced operator attention because of automation of the coagulant addition	Hours per day	1	0.5	0.5	hrs	365 days x "C" x $75/hr	$13,688
Benefit 3								
Benefit 4								
Benefit 5								
Benefit 6								

Part B - Changes in O&M Cost Calculations

	Description	Key Metric	Base period A	New B	Change C	Units D	Annual $Savings Formula E	Annual $Cost F
Cost 1	Instrum maintenance and calibration	Maintenance hrs/mnth	0	4	4	hrs/mnth	12 mnth x "C" x $80/hr	$3,840
Cost 2								
Cost 3								
Cost 4								
Cost 5								

Part C - Document Basis for Calculations

	Data Source	Comments
Benefit 1	Reduction in chemical use based on studies of similar agencies who have introduced coagulant feed automation and that have vabiale flow rates and raw water quality.	Reference studies by Dentel, Kinerery, 1988
Benefit 2	Reduction in hours is based on estimates by District Operators	
Benefit 3		
Benefit 4		
Benefit 5		
Benefit 6		
Cost 1	Estimate by District maintenance staff.	
Cost 2		
Cost 3		
Cost 4		
Cost 5		

STEP 5 - ANALYZE FINANCIAL BENEFITS

Cost-Benefit Analysis - Labor Savings Estimates

Instructions: Use the benefits worksheets to develop the basis or justification for the estimated benefits and allocate them to the year these are expected

		YEAR									
ITEM	NPV	1	2	3	4	5	6	7	8	9	10
Shift 1	$0										
Shift 2	$0										
Shift 3	$0										
Weekend	$0										
Travel/Over Time	$110,463	$13,687	$13,687	$13,687	$13,687	$13,687	$13,687	$13,687	$13,687	$13,687	$13,687
TOTAL	**$110,463**	**$13,687**	**$13,687**	**$13,687**	**$13,687**	**$13,687**	**$13,687**	**$13,687**	**$13,687**	**$13,687**	**$13,687**

Cost-Benefit Analysis - Energy Cost Savings

		YEAR									
ITEM	NPV	1	2	3	4	5	6	7	8	9	10
Raw Water Pumping	$0										
Flocculation	$0										
Filter Back Wash	$0										
High Service Pumping	$0										
Chemical Feed	$0										
Unit Process	$0										
Unit Process	$0										
Unit Process	$0										
Unit Process	$0										
TOTAL	**$0**	**$0**	**$0**	**$0**	**$0**	**$0**	**$0**	**$0**	**$0**	**$0**	**$0**

Cost-Benefit Analysis - Chemical Cost Savings

		YEAR									
ITEM	NPV	1	2	3	4	5	6	7	8	9	10
Alum	$58,916	$7,300	$7,300	$7,300	$7,300	$7,300	$7,300	$7,300	$7,300	$7,300	$7,300
Polymer	$0										
Filter aids	$0										
Chlorine	$0										
Ammonia	$0										
Process Chemical	$0										
Process Chemical	$0										
Process Chemical	$0										
Process Chemical	$0										
Process Chemical	$0										
TOTAL	**$58,916**	**$7,300**	**$7,300**	**$7,300**	**$7,300**	**$7,300**	**$7,300**	**$7,300**	**$7,300**	**$7,300**	**$7,300**

Cost-Benefit Analysis - O&M Savings

		YEAR									
ITEM	NPV	1	2	3	4	5	6	7	8	9	10
SCADA Maintenance	$0	$0	$0	$0	$0	$0	$0	$0	$0	$0	$0
Instrument Calibration	$0										
Reporting	$0										
Misc	$0										
Misc	$0										
TOTAL	**$0**	**$0**	**$0**	**$0**	**$0**	**$0**	**$0**	**$0**	**$0**	**$0**	**$0**

STEP 4 - DEVELOP PROJECT COSTS

Cost-Benefit Analysis Planning, Design and Engineering Costs

		YEAR									
ITEM	**NPV**	**1**	**2**	**3**	**4**	**5**	**6**	**7**	**8**	**9**	**10**
Planning											
As Built Documentation	$0										
Preliminary Design	$3,074	$3,200									
Master Planning	$0										
	$0										
Design											
Preliminary Design Document	$0										
Intermediate Design Submittal	$0										
Final Design Submittal	$4,323	$4,500									
Bid Set	$0										
Bidding Services	$480	$500									
Construction Phase											
Submittal Review	$0										
Administration	$0										
Final Design Submittal	$0										
Bid Set	$0										
Bidding Services	$480	$500									
Construction RE	$961	$1,000									
TOTAL	**$9,318**	**$9,700**	**$0**	**$0**	**$0**	**$0**	**$0**	**$0**	**$0**	**$0**	**$0**

STEP 4 - DEVELOP PROJECT COSTS

Cost-Benefit Analysis - Construction Costs

Instructions: This worksheet can be used to identify construction costs. The costing spreadheets in the Appendix can also be used to support more detailed estimates. Enter costs expected and the year the expenses are expected to occur.

		YEAR									
ITEM	**NPV**	**1**	**2**	**3**	**4**	**5**	**6**	**7**	**8**	**9**	**10**
Total Construction Cost From Costing Spreadsheets (If Used)	$40,394	$42,050									
SCADA Software	$0										
Printers & Peripherals	$0										
Engineering	$0										
Engineering	$0										
Control Panels	$0										
Panel Mounted Devices	$0										
HMI Configuration	$0										
PLC Programming	$0										
Misc	$0										
Field Devices	$0										
Equipment	$0										
Equipment	$0										
Equipment	$0										
Primary Instruments	$0										
Field wiring	$0										
Start Up	$0										
Training	$0										
Documenation	$0										
TOTAL	**$40,394**	**$42,050**	**$0**	**$0**	**$0**	**$0**	**$0**	**$0**	**$0**	**$0**	**$0**

* Construction cost estimate completed using the construction cost estimating tools provided in Appendix C.

STEP 4 - DEVELOP PROJECT COSTS

Cost Benefit Analysis - Post Acceptance Support O&M Costs

	YEAR									
NPV	**1**	**2**	**3**	**4**	**5**	**6**	**7**	**8**	**9**	**10**
$10,665	$0	$1,500	$1,500	$1,500	$1,500	$1,500	$1,500	$1,500	$1,500	$1,500
$3,840	$0	$498	$525	$550	$550	$550	$550	$550	$550	$550
$0										
$0										
$0										
$0										
$14,505	**$0**	**$1,998**	**$2,025**	**$2,050**	**$2,050**	**$2,050**	**$2,050**	**$2,050**	**$2,050**	**$2,050**

STEP 4 - DEVELOP PROJECT COSTS

ROXBOROUGH COAGULATION CONTROL UPGRADE PROJECT
OPERATIONS AND MANAGEMENT CHECKLIST

Operations and Maintenace Savings and Costs checklist

Document the annual costs before and/or after automation for the items listed below. Indicate the impact after the automation projects have taken place for the quantitative and qualitative items listed. Where the exact value is unknown, estimate to the

Before Automation	Annual Usage	Units	Annual Costs	Cost basis
Power				
Chemicals				
Operating Labor cost				
Control System Maintenance				
Other				
Other				
Other				
Other				

Document the annual costs after automation for the items listed below. Indicate the estimated savings in either $ or as a percentage, that may have resulted from the automation improvements or operational changes

After Automation	Annual Usage	Units	Annual Cost	Savings basis
Quantitative - Improved Conditions				
Power Costs				
Chemical Costs				
Operating Labor Costs				
Control System Maint. Costs				
Other				
Other				
Other				

STEP 5 - CALCULATE PROJECT NET PRESENT VALUE

COST BENEFIT ANALYSIS SUMMARY
ROXBOROUGH COAGULATION UPGRADE THEORETICAL EXAMPLE

Date of Estimate:	Jun-06	
Cost of Money:	7.1%	
Inflation Rate:	3.0%	
Effective Discount Rate	4.1%	
Project Life Cycle	10	years

		YEAR										
COST ITEM	**NPV**	**1**	**2**	**3**	**4**	**5**	**6**	**7**	**8**	**9**	**10**	**TOTAL**
Planning, Design, Engr	$9,318	$9,700	$0	$0	$0	$0	$0	$0	$0	$0	$0	$0
Construction	$40,394	$42,050	$0	$0	$0	$0	$0	$0	$0	$0	$0	$42,050
Post Acceptance Support	$14,220	$0	$2,000	$2,000	$2,000	$2,000	$2,000	$2,000	$2,000	$2,000	$2,000	$18,000
												$0
SUBTOTAL	**-$63,932**	**-$51,750**	**-$2,000**	**-$2,000**	**-$2,000**	**-$2,000**	**-$2,000**	**-$2,000**	**-$2,000**	**-$2,000**	**-$2,000**	**-$69,750**

		YEAR										
BENEFIT ITEM	**NPV**	**1**	**2**	**3**	**4**	**5**	**6**	**7**	**8**	**9**	**10**	**TOTAL**
Labor	$110,463	$13,687	$13,687	$13,687	$13,687	$13,687	$13,687	$13,687	$13,687	$13,687	$13,687	$136,870
Energy	$0	$0	$0	$0	$0	$0	$0	$0	$0	$0	$0	$0
Chemicals	$58,916	$7,300	$7,300	$7,300	$7,300	$7,300	$7,300	$7,300	$7,300	$7,300	$7,300	$73,000
Misc - Maint,	$0	$0	$0	$0	$0	$0	$0	$0	$0	$0	$0	$0
												$0
												$0
SUBTOTAL	**$169,379**	**$20,987**	**$20,987**	**$20,987**	**$20,987**	**$20,987**	**$20,987**	**$20,987**	**$20,987**	**$20,987**	**$20,987**	**$209,870**
TOTAL	**$105,447**	**-$30,763**	**$18,987**	**$18,987**	**$18,987**	**$18,987**	**$18,987**	**$18,987**	**$18,987**	**$18,987**	**$18,987**	**$140,120**

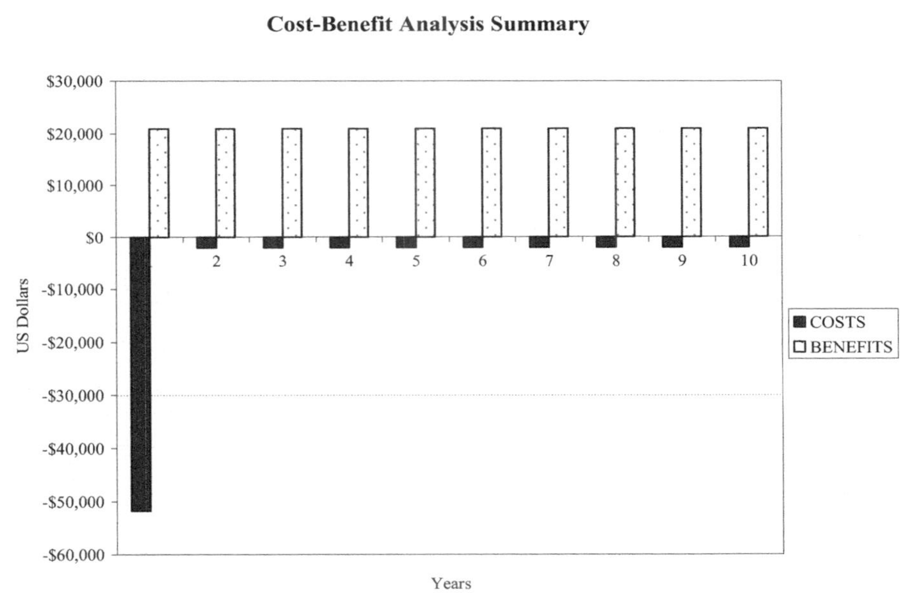

Figure A.2 Example NPV spreadsheet

APPENDIX B
CASE STUDIES

1. CITY OF HENDERSON, NV

Evaluation Date: March 29, 2006

Utility Participants: Michael Neher, Michael Morine, Frank Acito

Performed By: Dave Roberts, Alan Carrie

Project: Utility SCADA Upgrade

Utility Overview

The City of Henderson is the second largest city in the State of Nevada with a population of 250,000. The City operates a 15 mgd water treatment facility built in 1994 at a cost of $14 million. Water from the treatment plant and from Southern Nevada Water Authority is used to supply the City's water needs averaging 70 mgd. The City has Colorado River water rights to nearly 16,000 acre feet annually, taken from Lake Mead.

Project Description and Goals

In addition to providing high quality water, an important design objective for the new plant was to make it as economical to run as was possible. To this end the designers were required to use automation to maintain water quality and minimize labor costs. Automation would enable the plant to operate 24/7 but be manned for a single shift.

System Operation Description

The plant is attended by operators for 10 hours each day, 7 days a week. During the remaining 14 hours, the SCADA system alerts an on-call operator of abnormal conditions via phone, cell phone message (describing the problem) or by paging. A staff of 6 control system technicians and 2 electricians, is responsible for maintaining the SCADA system which integrates the water treatment plant, distribution system, the 32 mgd wastewater reclamation plant, and the collection system. Operations and maintenance staff perform maintenance duties as well as respond to the plant and distribution system operational needs.

Automation Features Description

The plant is controlled automatically using local PLC controllers (Square D) which communicate with a centralized SCADA system (CRISP HMI). During design a high level of automation was targeted with the aim of running the plant attended during the 10-hour day shift only. Automation included flow pacing of chemicals, and chlorine residual control. Filter backwashes were automated with the option of manual initiation. Alarms during the day are handled by in-plant operators. At night, an automatic dial-out system with voice annunciation notifies on-call staff.

Research

The City and designers did not perform a formal cost/benefit analysis to determine the level of automation. The approach was to provide a design which met the high water quality and efficiency goals set by Utility management. It was considered that private water operators ran the most efficient treatment plants and the City used their automation approach as a guideline.

Plant Schematic

The following diagram is a simplified schematic of the plant. UV disinfection was added after the original plant was built.

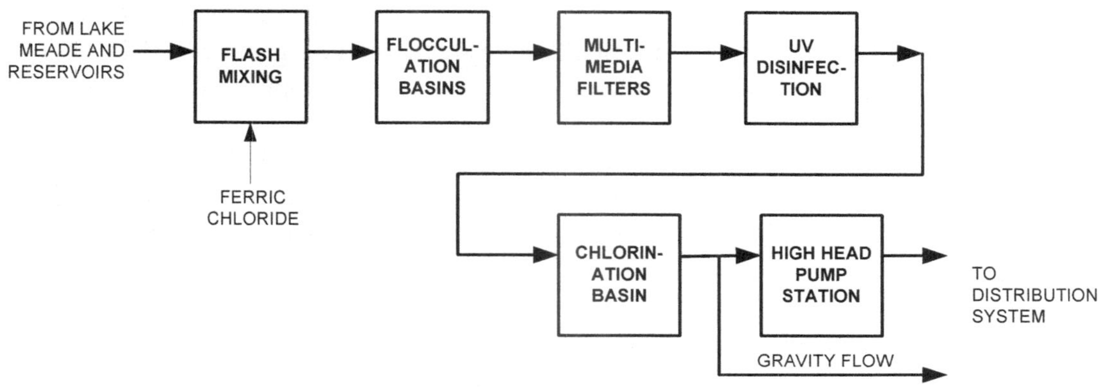

Figure B.1 Henderson process overview

Expected Benefits

Automation was expected to:

- Maintain a consistent high quality treated water output
- Minimize operational costs
- Minimize the plant maintenance costs

Project Construction Costs and Estimated Savings

The 1994 costs for the project were approximately:

Water treatment Plant	=	$ 10,733,000
SCADA System	=	$ 2,992,000

The SCADA system cost included a wastewater reclamation plant (10 mgd), Water treatment plant (15 mgd), and a number of remote pump stations and reservoirs, and collection system lift stations. Data records do not show the exact cost split between these projects. The utility staff estimated approximately one third of the SCADA costs are attributable to the water treatment plant.

No cost figures are available to show how much additional money was spent to implement a high level of automation, consistent with unattended operation for 14 hours a day. Approximately 50% of the instrumentation and control construction cost is attributable to implementing an automated system as opposed to a "monitoring only" solution. The construction cost increase could therefore be calculated as 0.33 x 0.5 x $2,992,000 = $494,000.

Project Outcome

The plant automation system has been running reliably since it was installed in 1994. Although a formal business case was not developed by the City to justify the cost of automation, staff believes the investment to be an unqualified success. It was felt by Operations that the original design goals of minimizing operating costs by investing in automation were achieved. The City is upgrading some parts of the SCADA system, including the SCADA historian, and will eventually upgrade the HMI software since this is no longer supported by the manufacturer. This upgrade will significantly improve data analysis and reporting.

If some assumptions are made concerning savings, then an NPV can be calculated using the known construction costs and the operational costs listed below. These calculations are hypothetical, but could be similar to those done to develop a business case at the pre-design stage of the project.

Reported Water Plant Operational Costs For 2005

Plant labor costs: The estimated plant labor costs for both operations and maintenance are estimated at approximately $235,000 yr. This cost includes a 40% burden.

Plant energy costs: Electrical and other energy charges are estimated at $383,600/yr.

Plant chemical costs: Estimated at $80,300/yr which includes chemicals for disinfection, coagulation, fluoridation, flocculation, and corrosion control. Chemicals were sodium hypochlorite (32%), polymer (15.5%), fluoride (31.5%), zinc orthophosphate, (8 %) and ferric chloride (13%). The percentages are of the total chemical cost.

Other costs: These include $3,560,000 for raw water and approximately $200,000 for maintenance materials.

Summary:

Chemical costs	=	$ 80,300
O&M labor costs	=	$ 235,000
Energy costs	=	$ 383, 600
Other costs	=	$ 3,756,000

Total operating cost =	$ 4,454,900	
Water production (MG) =		3,828
Total operating cost/MG =		$ 981

Calculating the Benefits of Automation

Because a formal NPV benefit analysis was not required when the project was being developed, a proper economic analysis cannot be presented. Never the less an analysis can be made based on probable savings. The following calculations are not based on data provided by the City, but rather on hypothetical chemical and operator labor savings which could easily have been generated by automating the treatment process. These calculations are provided solely to illustrate the NPV calculation methodology.

Savings

For the purpose of calculating a NPV, the following savings assumptions have been made:

- Labor savings due to automation are equivalent to one person. This could be made up of a reduction of operator labor of 1.5 persons and an increase in maintenance costs of 0.5 person equivalents.
- Chemical cost savings due to automation are estimated to be approximately 10 percent per year. In 2005, this would mean a savings of $8,207.
- Assume that labor and chemical costs have increased in cost an average of 4 percent over the past 10 years.
- The effective discount rate is 3 percent per year.
- Approximately one third of the SCADA construction costs are attributable to the water treatment plant.
- Approximately 50% of the instrumentation and control construction cost is attributable to implementing an automated system as opposed to a "monitoring only" solution. The construction cost increase would therefore be 0.33 x 0.5 x $2,992,000 = $ 494,000.

NPV calculation

Using the above assumptions, the NPV for a 10-year period is approximately $249,365. Hence the automation has a positive benefit over the 10 year period considered based on the assumptions made. This does not account for the indirect benefits stated in the project goals such as water quality and consistency improvements, which add considerably to the value of the project. Note that the NPV calculations use the chemical and labor costs/savings estimated for 1994, the year the plant was built.

Cost-Benefit Analysis Summary
Henderson SCADA

Date of Estimate:	Jun-06	
Cost of Money:	7%	
Inflation Rate:	4%	
Effective Discount Rate	3.0%	
Project Life Cycle	10	years

		YEAR										
COST ITEM	**NPV**	**1**	**2**	**3**	**4**	**5**	**6**	**7**	**8**	**9**	**10**	**TOTAL**
Planning, Design, Engr	$0	$0	$0	$0	$0	$0	$0	$0	$0	$0	$0	$0
Construction	$479,612	$494,000	$0	$0	$0	$0	$0	$0	$0	$0	$0	$494,000
Post Acceptance Support	$15,119	$0	$2,000	$2,000	$2,000	$2,000	$2,000	$2,000	$2,000	$2,000	$2,000	$18,000
												$0
SUBTOTAL	**-$494,730**	**-$494,000**	**-$2,000**	**-$2,000**	**-$2,000**	**-$2,000**	**-$2,000**	**-$2,000**	**-$2,000**	**-$2,000**	**-$2,000**	**-$512,000**

		YEAR										
BENEFIT ITEM	**NPV**	**1**	**2**	**3**	**4**	**5**	**6**	**7**	**8**	**9**	**10**	**TOTAL**
Labor	$680,194	$0	$87,360	$87,360	$87,360	$87,360	$87,360	$87,360	$87,360	$87,360	$87,360	$786,240
Energy	$0	$0	$0	$0	$0	$0	$0	$0	$0	$0	$0	$0
Chemicals	$63,901	$0	$8,207	$8,207	$8,207	$8,207	$8,207	$8,207	$8,207	$8,207	$8,207	$73,863
Misc - Maint,	$0	$0	$0	$0	$0	$0	$0	$0	$0	$0	$0	$0
												$0
												$0
SUBTOTAL	**$744,095**	**$0**	**$95,567**	**$95,567**	**$95,567**	**$95,567**	**$95,567**	**$95,567**	**$95,567**	**$95,567**	**$95,567**	**$860,103**
TOTAL	**$249,365**	**-$494,000**	**$93,567**	**$93,567**	**$93,567**	**$93,567**	**$93,567**	**$93,567**	**$93,567**	**$93,567**	**$93,567**	**$348,103**

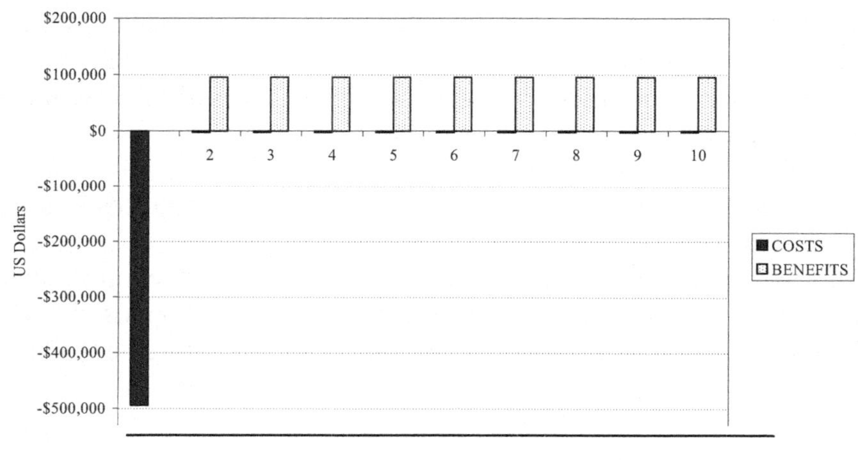

Figure B.2 Henderson NPV analysis

2. ONONDAGA COUNTY, MARCELLUS, NY

Evaluation Date: March 10, 2006

Utility Participants: Nicholas Kochan

Performed By: Dave Roberts, Dave Kubel

Project: Otisco Lake Water Treatment Plant

Utility Overview

The Onondaga County Water Authority (OCWA) provides treated water to approximately 375,000 people. Water is obtained from 3 sources: water is drawn from Lake Otisco and treated by the utility at the Otisco Lake WTP; water is purchased from the Metropolitan Water Board and does not require further treatment; and a small portion of water is purchased from the City of Syracuse Water Department and does not require further treatment.

In 2004, approximately 17.5 million gallons per day or 43% of OCWA's water came from Otisco Lake. Water from the lake is the preferred source, since it is less expensive. Both treated and purchased water sources are required to meet the system demand. The treatment plant has 3 hours of storage in the clearwells, but cannot be shut down for a longer period than that without the risk of draining the distribution system. The water from the lake is gravity fed to the WTP, gravity fed through the plant, and gravity fed to the distribution system. Booster pumping is required in the distribution system.

A SCADA system monitors and controls approximately 70 sites, including the WTP. The operations center is approximately 30 miles from the plant and provides the SCADA monitoring location. The SCADA system provides remote alarming, with an autodialer at the WTP as a backup. Total SCADA System I/O is approximately 700 points.

Project Description and Goals

- Initial Service Date: 1986
- Gallons Per Day: 20 million
- Population Served: 146,200
- Million Gallons treated per year: 6,293
- Treatment Process: Direct Filtration

OCWA has two intake pipes in Otisco Lake. The water entering these pipes is immediately chlorinated to provide disinfection and to discourage the growth of zebra mussels. The water travels, by gravity, approximately 5 miles to OCWA's Water Treatment Plant in Marcellus, NY. The WTP, a direct filtration plant, was constructed in 1986 with a design capacity of 20 mgd. Water first enters the rapid mix tank where a coagulant (polyaluminum chloride - PACL) and a taste and odor control chemical (powdered activated carbon) is added. After 30 seconds of mixing, the water enters the contact basins where some of the particles settle and are cleaned out later.

The contact time in these basins also allow the powdered activated carbon (used about 4 months of the year) to adsorb organic taste and odor causing chemicals. After about 1 hour of contact time, the water enters the filters. Particles are removed as the water passes through one of four multimedia filters. A non-ionic polymer is added as a filter aid in very low doses during times when filter breakthrough occurs. These filters consist of granular activated carbon, silica-sand, hi-density sand, and three layers of gravel. The filters are washed periodically and the water used to do this is collected in lagoons and allowed to settle. The clear water is recycled back to the start of the treatment plant to be treated again. After filtration, the water is again disinfected with chlorine and fluoride is added.

The water is stored in large tanks at the treatment plant to provide adequate contact time for the chlorine to work. Once the water leaves the tanks orthophosphate is added to provide a coating for the pipes in the distribution system and in the homes, to prevent the leaching of lead and copper from the pipes into the water.

System Operation Description

The plant is operated in an unattended manner eight hours per day. During this time it is monitored via SCADA from the operations control center. Prior to September 11, 2001, the operator at the control center would leave the SCADA system occasionally to perform various tasks in the distribution system. Since then, the SCADA system is monitored 24/7 by an operator. Response time to an incident is approximately 30 minutes during unmanned operation.

Staffing hours are not tracked separately from an accounting standpoint between the plant and the distribution system. Estimates were made of the current staffing during the interviews. Instrumentation and Controls maintenance equipment and supplies are budgeted for $5K-$10K/yr. They have three instrument techs for the plant and distribution system. They have a maintenance tech at the plant. The plant utilizes approximately 1/2 FTE for maintenance purposes.

Operations staffing is 16 hours per day with 1 FTE on-site, plus a shared portion of an FTE during off hours (operator watching SCADA for distribution and treatment plant). Operators average approximately $70K-$75K/yr with overhead included.

Automation Features Description

The plant includes Bristol RTUs. The filter backwash has an automated program, but this is performed manually, as the operators are able to control it better manually. Communications from the plant to operations center is over an ISDN line. In the event of a loss of communications, the plant is attended.

The plant treats the water to ≤ 0.10 NTU. Due to this constraint, there are times when the plant flow is reduced to maintain this effluent water quality. The Utility is considering process modifications to be able to maintain the 20 mgd flow rate at all times. The plant has operated in a partially unattended manner for 20 years. The project being considered in this case study is the initial plant construction, and various modifications over the years. There are no state regulations that prevent them from operating unattended.

In general, the plant is operated in a manual mode. This includes initiating backwashes, and filling the backwash tank. The SCADA system will alarm the operator in the event of a problem at the plant. An auto dialer at the plant is used as a back-up to this and will call a separate operator. This ensures that, in the event something were to happen to the operator at the SCADA system, another operator will be able to respond.

In the 20 years of operating the plant unattended, there have been no problems that have resulted in the need to shut down the plant. Shut down would be a significant problem due to the potential vacuum created if the plant is offline long enough to drain the clear well.

Of significant operational consideration is the raw water turbidity. The lake will go through several temperature inversions each year, resulting in high turbidity. During these events, staff will attend the plant overnight and reduce the plant production. They will also attend the plant overnight due to algae issues at times.

Remote control of the plant allows remotely adjusting the Raw Water Flow Control Valve only. This valve was added after 7 years of unattended operation, allowing for better remote control. This allows remotely backing off the flow during a turbidity event. Plant effluent flow is also remotely controlled, but it is controlled indirectly by controlling the distribution system.

Research

The focus of the Utility team and designers was to provide a design which meets the high water quality and efficiency goals set by Utility management. As a result, a formal cost/benefit analysis was not used to determine the level of automation.

Plant Process Schematic

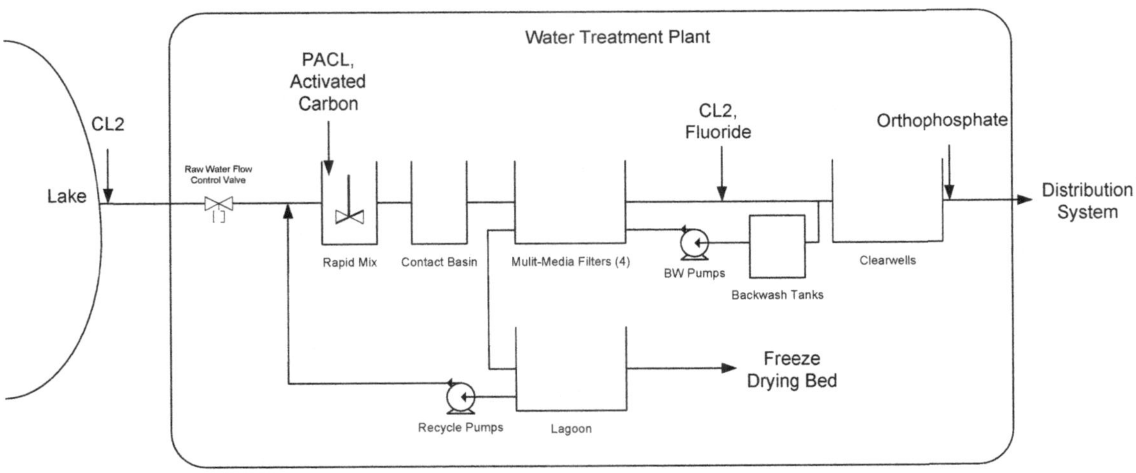

Figure B.3 Simplified Otisco Lake process schematic

Expected Benefits

Automation was expected to:

- Maintain a consistent high quality treated water output.
- Minimize operational costs
- Minimize the plant maintenance costs

Project Construction Costs and Estimated Savings

The plant controls were largely installed with the plant in 1986. Some changes were made since then, including the addition of the raw water flow control valve. The plant was constructed for approximately $6M in 1986. No cost figures are available to show how much additional money was spent to implement automation necessary for unattended operation for 8 hours a day.

Project Outcome

The plant automation system has been running reliably since it was installed in 1986. Although a formal business case was not developed, to justify the cost of automation, staff believes the investment to be a success and that the high level of automation is essential for operating the plant. It was felt by Operations that the original design goal of minimizing operating costs by investing in automation was achieved.

3. PLACER COUNTY WATER AGENCY, AUBURN, CA

Evaluation Date: March 13, 2006

Utility Participants: Wally Cable with cost information provided by Brent Smith and Tony Firenzi

Performed By: Dave Roberts, Dave Kubel, Dean Schoeder

Project: Alta Water Treatment Plant

Utility Overview

The Placer County Water Agency operates eight individual treated water systems. These water systems include Alta, Applegate, Bianchi, Auburn/Bowman, Colfax, Foothill-Sunset, Lahontan and Monte Vista. Six of the water systems are supplied through water treatment plants that treat surface water supplied via the PCWA canal system. The Bianchi system serves surface water purchased from the City of Roseville. The Lahontan system is supplied by wells. All of the treatment plants, ranging up to 55 mgd, run in an unattended manner for at least a portion of each day.

The Agency operates an extensive raw water distribution system that includes 165 miles of canals, ditches, flumes and several small reservoirs. A significant amount of Agency raw water irrigates agricultural land and golf courses. Drinking water is produced through a network of eight water treatment plants. More than 150,000 people depend on PCWA water supplies.

The Agency is governed by a five-member Board of Directors, elected to four-year terms by voters residing within five geographical districts of Placer County. The Board of Directors meets twice monthly in regular session and holds special meetings as needed. The Board employs a General Manager to administer all Agency activities, services and employment, and a General Counsel to advise the Agency on legal and regulatory matters. The Agency employs approximately 180.

Project Description and Goals

Alta is an existing plant that provides an average of 200+ GPM of finished water. Automatic monitoring and control functions have been incrementally installed over the years. The plant was initially set up to be operated locally with only a storage level telemetry link to PCWA Water Treatment headquarters. PCWA decided to install streaming current controllers to automatically adjust chemical dosage to address water quality shifts caused by wide ranging raw water turbidity levels. When the new turbidity reporting regulations went into effect, there was a need to install a turbidimeter on the outlet of each of the three filters. The plant was next included in an enhanced SCADA system where much greater remote monitoring was implemented. Recently, Alta WTP was included in the system-wide SCADA upgrade with view and control anywhere abilities. The identical control screens are now available for use at every other treatment plant and at strategic other locations in the Agency.

System Operation Description

The plant automation was set up to operate unattended. Roving operators stop at the plant once a day to check status of equipment, etc. Turbidimeters on the plant influent and on the filter effluent shut down the raw water pumps on high turbidity. Prior to upgrading the automation pump shutdowns were not communicated to the PCWA operations headquarters. Pump shutdowns were previously not known until the finished water storage tank level went low. The finished water low tank level alarm is PCWA operations headquarters.

The control system hardware at the plant is PLC based. The plant control system performs an automated backwash on each filter every 3 days. Chemical feed pumps are automatically controlled. The system monitors pH, turbidity, and chlorine. The local control system includes an autodialer to advise staff of alarms requiring attention. Staff requires an hour to reach the plant to attend to these situations. The on-call operator also has a laptop and can dial in to view and control any of the eight PCWA treatment plants.

Expected Benefits

Reason	Expected benefit
Need to comply with reporting requirements	More accurate, reliable data collection
Ability to operate plant under wider range of raw water conditions	Fewer plant shutdowns caused by "poor" raw water quality
Automated control of treatment process	Rapid response to plant upsets
Utilize hardware design standards	Improve consistency of equipment to reduce design and maintenance costs
Utilize software design standards	Reuse programs from prior projects to improve consistency of software reduce construction and maintenance costs

Project Costs and Estimated Savings

The following provides a NPV estimate of the automation improvements at the Alta WTP. Since the automation improvements were done as a part of an overall project to provide monitoring and control at several sites the estimate tried to identify the costs that were related to providing a high level of automation at the Alta plant by itself. The analysis indicates a favorable return on investment an over the 10 year project time frame shows a NPV of approximately $46,252.

COST BENFIT ANALYSIS SUMMARY
PCWA - ALTA WTP AUTOMATION

Date of Estimate:	Apr-06	
Cost of Money:	7.1%	
Inflation Rate:	3.0%	
Effective Discount Rate	4.1%	
Project Life Cycle	10	years

		YEAR										
COST ITEM	**NPV**	**1**	**2**	**3**	**4**	**5**	**6**	**7**	**8**	**9**	**10**	**TOTAL**
Planning, Design, Engr	$25,357	$24,254	$2,143	$0	$0	$0	$0	$0	$0	$0	$0	$0
Construction	$139,553	$0	$0	$145,274	$0	$0	$0	$0	$0	$0	$0	$145,274
Maintenance/Support	$15,468	$0	$0	$2,500	$2,500	$2,500	$2,500	$2,500	$2,500	$2,500	$2,500	$20,000
												$0
SUBTOTAL	**-$180,378**	**-$24,254**	**-$2,143**	**-$147,774**	**-$2,500**	**-$2,500**	**-$2,500**	**-$2,500**	**-$2,500**	**-$2,500**	**-$2,500**	**-$191,671**

		YEAR										
BENEFIT ITEM	**NPV**	**1**	**2**	**3**	**4**	**5**	**6**	**7**	**8**	**9**	**10**	**TOTAL**
Labor	$226,630	$0	$0	$33,800	$33,800	$33,800	$33,800	$33,800	$33,800	$33,800	$33,800	$270,400
Energy	$0	$0	$0	$0	$0	$0	$0	$0	$0	$0	$0	$0
Chemicals	$0	$0	$0	$0	$0	$0	$0	$0	$0	$0	$0	$0
Misc - Maint,	$0	$0	$0	$0	$0	$0	$0	$0	$0	$0	$0	$0
												$0
												$0
SUBTOTAL	**$226,630**	**$0**	**$0**	**$33,800**	**$33,800**	**$33,800**	**$33,800**	**$33,800**	**$33,800**	**$33,800**	**$33,800**	**$270,400**
TOTAL	**$46,252**	**-$24,254**	**-$2,143**	**-$113,974**	**$31,300**	**$31,300**	**$31,300**	**$31,300**	**$31,300**	**$31,300**	**$31,300**	**$78,729**

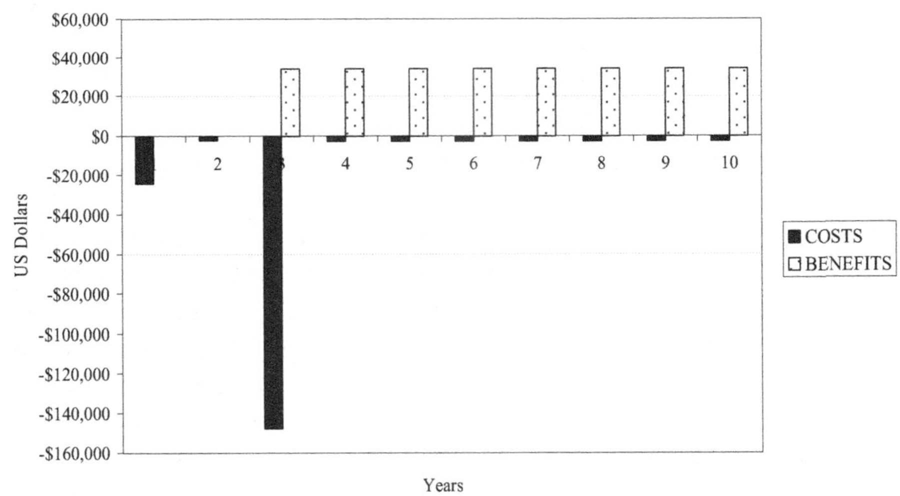

Cost-Benefit Analysis Summary

Figure B.4 PCWA Alta WTP NPV analysis

4. ARIZONA - AMERICAN WATER, ANTHEM, AZ

Evaluation Date: May 2, 2006

Utility Participants: Mike Helton, Jeff Marlow, Jim Grooman

Performed By: Dave Roberts, Shawn Rohr

Project: Anthem Water Campus

Utility Overview

American Water was founded in 1886 as the American Water Works & Guarantee Company and reorganized in 1947 as American Water Works Company, Inc. On January 10, 2003 the company was acquired by RWE, was renamed "American Water," and became a part of RWE's water division. The majority of the company's activities are centered in locally managed utility subsidiaries that are regulated by the state in which each operates. These state utilities are supported by the resources of American Water and are an integral part of the communities they serve.

The company also owns subsidiaries that manage municipal water and wastewater systems under contract and others that supply homeowners, businesses, and communities with water-resource-management products and services.

Project Description and Goals

- Initial Service Date: 1999
- Gallons Per Day: 7 million
- Population Served: 13,000
- Small Footprint (approximately 10 acres)
- Immersed Membrane Filtration Technology

The Anthem Water Treatment Plant (WTP) receives water from the Waddell Canal, which connects the Central Arizona Project Canal and Lake Pleasant. Surface Water from the Waddell Canal is drawn through a trash rack and pumped to the Anthem Water Campus through a 30-inch diameter pipeline approximately 8.8 miles long. At the Water Campus, the flow is directed into the raw water storage reservoir. From the storage reservoir, the water enters the raw water pump station wet well where submersible pumps convey the water through a strainer and influent meter to the membrane filtration (MF) process tanks.

The MF system has a maximum design capacity of 7.4 million gallons per day and utilizes a membrane filtration process. After treatment by the MF system, the permeate pumps deliver the water through a treated water meter and into two finished water storage reservoirs. An ultraviolet disinfection system, followed by chlorine contact tanks, serves as the primary disinfection process for the water treatment plant.

An interesting aspect to the Anthem facility is that it is a combined water and wastewater treatment plant. Both plants are operated by the plant staff.

In addition to providing high quality water, an important design objective for the new plant was to make it as economical to run as was possible. To this end, the designers were required to use automation to maintain water quality and minimize labor costs. Automation enabled the plant to operate 24/7 but be attended by operators for a single shift.

System Operation Description

The Campus is normally attended by four (4) operators that work both water and wastewater plants. Each operator works four (4) ten hour shifts per week. Operators are on site 7 days per week and the facility is attended for 10 hours per day. The number of operators on site in a given day varies but typically, there are a maximum of three operators on site on a given day. Overlap of work schedules can result in up to five operators on site on Wednesdays. Weekends alternate from two to three operators on-site.

During the remaining 14 hours, the SCADA system alerts an on-call operator of abnormal conditions via phone, cell phone message (describing the problem) or by paging.

When the plant was initially constructed, the plant was attended 7 days per week and operators were on-site for two (2) 8-hour shifts. Automation has aided in the ability to reduce the amount that the plant is attended.

A maintenance group supports the automation and there are plans to add an additional instrument technician to support the automation.

Automation Features Description

The plant is controlled automatically using local PLC controllers (Allen-Bradley) which communicate with a centralized SCADA system (Wonderware HMI). The SCADA system is programmed so that all the control logic is distributed in the PLCs and the SCADA HMI operates in a true supervisory mode. The SCADA computers are arranged in a dual redundant mode and are mirrored to provide back up in the event the primary computer fails.

Expected Benefits

The level of automation was not determined by using a formal cost/benefit analysis. The approach was to provide a design, which meets the high water quality, and efficiency goals set by Utility management. As a private water company, a key business driver is to run their plants in a highly efficient and cost effective manner.

Automation was expected to:

- Maintain a consistent high quality treated water output
- Minimize operational costs
- Minimize the plant maintenance costs

Project Construction Costs

The SCADA system cost included a wastewater treatment and reuse plant (3 mgd), Water treatment plant (7.4 mgd) a number of remote pump stations and reservoirs, and collection system lift stations. A detailed breakdown of construction costs was not readily available to show how much was spent to implement a high level of automation, consistent with unattended operation for 14 hours a day.

Estimating the Benefits of Automation

The plant automation system has been running reliably since it was installed in 1999. Although a formal business case was not developed to justify the cost of automation, staff believes the investment to be an unqualified success and that the high level of automation is essential for operating the plant. It was felt by Operations that the original design goals of minimizing operating costs by investing in automation were achieved.

Although a formal NPV analysis was not completed, the high level of automation has resulted in significant savings and has enabled the Utility to be able to run the facility in an unattended mode for a significant amount of time. This has resulted in significant labor savings. They have also realized some secondary tangible and intangible benefits including:

- A great deal of historical data is collected on the SCADA system that can be used to analyze the efficiency and performance factors for the plant. By analyzing some of this data, they were able to determine that an overflow between tanks was resulting in double the amount of pumping on one of the recycle streams. Fixing the overflow problem reduced energy costs.
- The enhanced pumping controls have allowed them to eliminate the purchase of supplemental water due to optimizing their pumping strategies. One of their customers (Golf Course) takes all of their water at once. Due to the enhanced pumping strategies implemented on the SCADA system, they avoid purchasing 150-acre feet of supplemental water per year at $83/acre foot, resulting in savings of approximately $12,450 per year.
- Some of the operators were initially intimidated by the high level of automation and computerization but once they were trained and understood how it could be used, the operators liked it. They have also have found that it makes them a more valuable employee if they have those computer skills. The operators believe that their jobs are easier and the scope of the operators' role has evolved. They can spend more time on evaluating data and operational parameters and not spend as much time on the repetitive activities that they may have without a high level of automation. The result is that the operator position is becoming a more skilled position and a morale boost that is embraced by the staff.
- The town of Anthem did not exist 8 years ago and the plant and the level of automation has enabled it to grow and serve the community rapidly.
- They use this plant as a showcase to show their efficiency and automation plays a key role in that. As a private utility, one of the services they offer is contract operations. This plant is used to demonstrate their capabilities.

5. IRVINE RANCH WATER DISTRICT, IRVINE, CA

Evaluation Date: March 10, 2006

Utility Participants: Carl Spangenberg, Dave Mazzarella

Performed By: Dave Roberts, Alan Carrie

Project: Deep Aquifer Treatment System

Utility Overview

Irvine Ranch Water District was established in 1961 as a California Water District that encompasses 133 square miles in southern central Orange County. It serves the city of Irvine and portions of Tustin, Newport Beach, Costa Mesa, Orange and Lake Forest. In 1997, Irvine Ranch Water District began providing water to the Santa Ana Heights community. The Carpenter Irrigation District chose to consolidate with IRWD in 2000, and at the beginning of 2001 a consolidation between IRWD and Los Alisos Water District of Lake Forest was completed.

Irvine Ranch Water District extends from the Pacific Coast to the foothills, with elevations ranging from sea level to 1,700 feet. The area served by IRWD is a semi-arid region with a mild climate and an average annual rainfall of 12-13 inches.

Project Description and Goals

Irvine Ranch Water District's Deep Aquifer Treatment System (DATS) purifies drinking water from the lower aquifer of the Orange County Groundwater Basin. Although the water from this aquifer is very high in quality, it was previously unusable caused by a brownish tint imparted from the remains of decaying ancient vegetation. However, new purification technologies make it possible and cost-effective to remove the color from this water. The DATS facility went on-line in early 2002.

The facility includes two wells that pump water from approximately 2000 ft. below ground level. This colored water enters the treatment plant and travels through tightly-wound membranes. Because the color molecules are much larger than the water molecules, they can readily be removed by the phenomenon defined as "size exclusion.". After undergoing this "nano-filtration" treatment, the water is clear. It then travels through degasifiers that remove low levels of methane gas and is disinfected with free chlorine prior to discharge to the wellfield transmission main.

The waste concentrate is discharged into the sanitary sewer system. The treatment process has a 92% efficiency rate. This means for every 100 gallons of colored water that passes through the facility, 92 gallons of clear water are produced and 8 gallons of concentrate. An engineering team is currently working on system upgrades that may be able to increase the efficiency as high as 98%. The DATS facility can treat up to 7.4 million gallons of water per day.

This project benefits all groundwater users in northern and central Orange County because it helps prevent wells from being affected by seepage of colored water into the middle aquifer used for drinking water by many cities and water districts.

System Operation Description

The plant is operated unmanned with control and monitoring from the Michelson Wastewater Treatment Plant 8 miles away. An Operator visits the facility for two hours each day to perform water quality bench tests, check operations and chemical use.

After color removal, the treated DATS water is combined with water from other wells, disinfected with chloramines, and sent to the distribution system.

SIMPLIFIED SCHEMATIC – DATS PROCESS

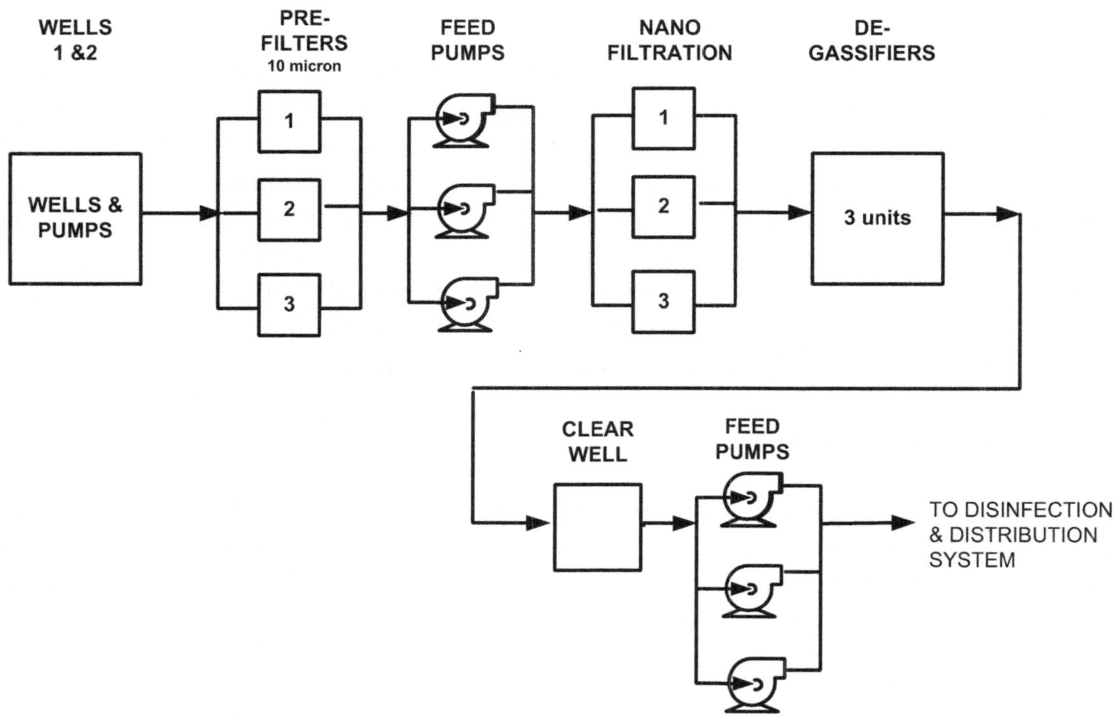

Figure B.5 IRWD process schematic

Automation Features Description

The plant is controlled automatically using a local PLC controller (Modicon) and a SCADA HMI (Wonderware). Operator control and monitoring can be from a workstation at the DATS facility or remotely from the nearby wastewater treatment plant.

Research

The District performed a cost/benefit analysis which included a comparison between a highly automated facility and one that was manually operated. This indicated that a high level of automation was justified. The District estimated that without automation the facility would require two operators and three shifts. During the project concept phase (1998) a cost comparison was made between automated and manual operation which showed automated operation would be the most economic approach. The cost figures were unavailable for inclusion in this report.

Expected Benefits

Automation was expected to:

- Maintain a consistent high quality treated water output
- Minimize operational costs

The 2001 costs for the project were approximately :

DATS treatment Plant	=	$ 10.5 million (excludes wells)
DATS SCADA System	=	Not Available

An important design goal for the DATS automation was minimization of operational costs. No cost figures were available to show how much additional money was spent to implement a high level of automation, consistent with unattended operation.

For the purpose of analysis, some typical costs were estimated. The SCADA system cost was estimated at approximately 10 % of the construction cost, and that 50% of the instrumentation and control cost is attributable to implementing an automated system as opposed to a "monitoring only" solution. Using these assumed figures the cost of advanced automation = 0.5 x 0.1 x $10.5 = $ 525,000.

Project Outcome

Operators currently spend 2 hours a day at the facility performing manual sampling and testing. This provides data needed for regulatory reports. The original design included an on-line instrument for color analysis but this proved unreliable. Color testing is now manual. A study is in progress looking at a replacement instruments that may prove to be more acceptable.

Water Plant Operational Costs For 2005

Plant Labor: No labor costs information was available, hence only a hypothetical NPV benefit can be calculated. If an assumed operator cost of $65/hour is used the difference between the cost of a fully manned and unmanned plant can be calculated.

Fully attended facility:

Number of Operators	= 6
Operator hourly cost	= $ 65/hr
Annual cost	= 6 x $65 x 52 weeks x 56 hrs/week = $ 1,135,680

Fully automated facility

Number of Operators = 1 full time operator that may spend only 2 hrs/day at the facility.

Operator hourly cost = $ 65/hr

Annual cost = 1 x $65 x 52 weeks x 40 hrs/week = $ 135,200

Annual maintenance costs associated with the automation, are estimated at $34,500 per year after the facility is fully commissioned.

Calculating the Benefits of Automation

The following calculations are based on the above hypothetical data to illustrate a simplified NPV calculation considering the estimated labor, construction cost and maintenance costs associated with the enhanced DATS automation.

Using the above assumptions the NPV for a 10 year period is approximately $6,060,289. Hence the automation has a significant positive benefit over the 10 year period considered based on the assumptions made and the avoided operations costs. This does not account for the indirect benefits stated in the project goals such as water quality, consistency improvements, and avoiding contaminating upper aquifers, which add considerably to the value of the project.

STEP 5 - CALCULATE PROJECT NET PRESENT VALUE

COST BENFIT ANALYSIS SUMMARY
IRWD DATS

Date of Estimate:	Mar-06	
Cost of Money:	7.1%	
Inflation Rate:	3.0%	
Effective Discount Rate	4.1%	
Project Life Cycle	10	years

		YEAR										
COST ITEM	**NPV**	**1**	**2**	**3**	**4**	**5**	**6**	**7**	**8**	**9**	**10**	**TOTAL**
Planning, Design, Engr	$110,951	$63,000	$52,500	$0	$0	$0	$0	$0	$0	$0	$0	$0
Construction	$504,323	$0	$525,000	$0	$0	$0	$0	$0	$0	$0	$0	$525,000
Post Acceptance Support	$222,213	$0	$0	$34,500	$34,500	$34,500	$34,500	$34,500	$34,500	$34,500	$34,500	$276,000
												$0
SUBTOTAL	**-$837,487**	**-$63,000**	**-$577,500**	**-$34,500**	**-$34,500**	**-$34,500**	**-$34,500**	**-$34,500**	**-$34,500**	**-$34,500**	**-$34,500**	**-$916,500**

		YEAR										
BENEFIT ITEM	**NPV**	**1**	**2**	**3**	**4**	**5**	**6**	**7**	**8**	**9**	**10**	**TOTAL**
Labor	$6,848,829	$0	$0	$1,021,446	$1,021,446	$1,021,446	$1,021,446	$1,021,446	$1,021,446	$1,021,446	$1,021,446	$8,171,571
Energy	$0	$0	$0	$0	$0	$0	$0	$0	$0	$0	$0	$0
Chemicals	$48,947	$0	$0	$7,300	$7,300	$7,300	$7,300	$7,300	$7,300	$7,300	$7,300	$58,400
Misc - Maint,	$0	$0	$0	$0	$0	$0	$0	$0	$0	$0	$0	$0
												$0
												$0
SUBTOTAL	**$6,897,776**	**$0**	**$0**	**$1,028,746**	**$1,028,746**	**$1,028,746**	**$1,028,746**	**$1,028,746**	**$1,028,746**	**$1,028,746**	**$1,028,746**	**$8,229,971**
TOTAL	**$6,060,289**	**-$63,000**	**-$577,500**	**$994,246**	**$994,246**	**$994,246**	**$994,246**	**$994,246**	**$994,246**	**$994,246**	**$994,246**	**$7,313,471**

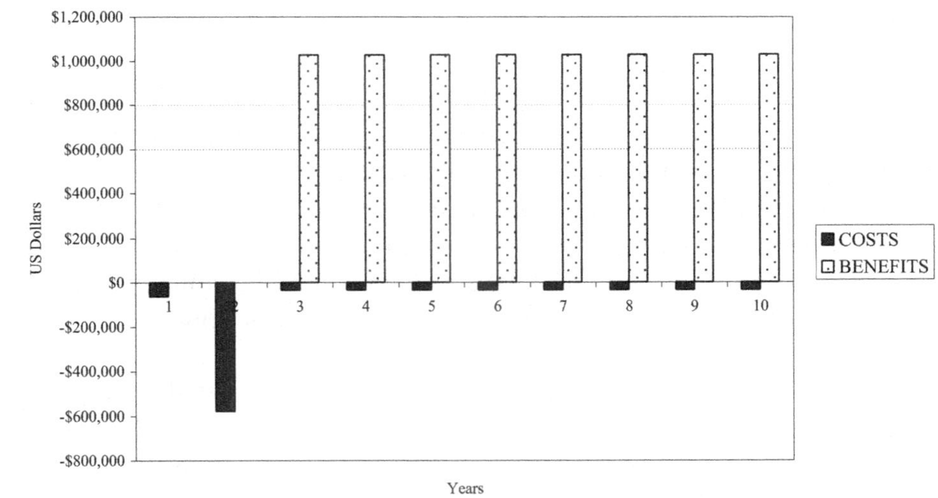

Cost-Benefit Analysis Summary

Figure B.6 IRWD NPV analysis

6. MEDFORD WATER COMMISSION, MEDFORD, OR

Evaluation Date: Initial data collected April 26, 2005

Utility Participants: Jim Stockton

Performed By: Dave Roberts

Project: Duff Water Treatment Plant

Utility Overview

The Medford Water Commission operates and maintains the water system that delivers high-quality drinking water to around 125,000 Rogue Valley residents. The Commission was established through a change in the City's Charter in 1922. Big Butte Springs is the Water Commission's primary water source, with the Rogue River used as a supplement during the summer months. Water is withdrawn at the Robert A. Duff Water Treatment Plant (Duff WTP) near TouVelle State Park.

The existing Duff WTP Plant is a conventional treatment facility that was built in 1968 with an original capacity of 15 mgd and was expanded in 1983 to 30 mgd and again in 2000 to 45 mgd. The eventual plant capacity of the facility was designed to be 65 MGD. It is currently utilized only during summer high demand periods. The summertime raw water characteristics are stable and of high quality with the water quality of the finished water with an average effluent turbidity of 0.05 NTUs.

Process Overview

The raw water supply is from the Rogue River and is provided with intake screenings for the removal of large materials. The screened raw water is then pumped using constant speed pumps. The influent flow is measured using a magnetic flow meter.

The treatment plant has chemical metering equipment for the treatment of the raw water including: chlorine, alum, polymer (two types), lime and activated carbon. The addition of these chemicals is paced on influent flow. The water then enters the contact basins where rapid mixing and flocculation occurs. The mixing of flocculating agent is done using hydraulic action at the inlet of the contact basins.

There are twelve constant rate filters that discharge treated water into a 5.0 MG storage reservoir. The filters are provided with backwash controls that include filter-to-waste capability. Backwash water is supplied to the filters with one backwash pump that draws treated water from the storage reservoir. Backwash water and filter to waste flows are sent to the Backwash Settling Lagoons 1 & 2. Settled water from the lagoons is returned to the river.

The treated water is chlorinated and five high service pumps are used to move the finished water into the distribution system. Distribution system storage is remote from the Duff WTP. Four of the five pumps are constant speed and the fifth is a variable frequency driven pump. The speed of the pump is controlled to maintain a pressure setpoint in the distribution system header. The variable speed pump can be controlled remotely from an existing control system that telemeters (on/off) control signals from the Service Center to the Duff WTP using a separate radio based telemetry system.

Project Description

During the period from 1998 to 2000 the automation systems at the Duff plant were upgraded. The following provides a summary of the automation improvements:

- An Allen-Bradley SLC PLC and iFIX Supervisory Control and Data Acquisition (SCADA) system replaced the existing manual controls and provides for a high level of process automation. The SCADA system includes two Workstations in the main control room, one Operator Interface Station (OIS) in the main control panel, one administrative and one laboratory workstation connected via an Ethernet network. The workstations connect to the PLCs via a DH+ network. The system includes approximately 11 PLCs, 1,200 field I/O points and 5,400 database points. The workstations provide an interface to the process and provide the ability to produce reports.
- Existing manual only backwash controls were replaced with PLC based control logic that allows for automatic, as well as, manual backwashing of the filters and plant control.
- All pneumatic control signaling devices including differential pressure sensors, level sensors, flow sensors and filter effluent discharge vale actuators were replaced with electronic 4-20 milliamp based devices.
- Hardwired backup controls were provided using local filter control panels in the filter gallery on the first floor.
- Met-One Particle counters were installed on all filter effluent lines and integrated with the plant SCADA system for historical data collection and plant control.
- Installed data collection and historical database applications linked to the SCADA system. This provides for daily, monthly and long term reports.
  Turbidimeters were installed on all filter effluent lines.
- Streaming current control to adjust coagulant dosage
- Instrumentation was installed in selected areas of the plant to improve the ability to monitor and control the plant including:
  - Flow switches on the discharge lines of the alum and polymer feed pumps for positive indication of chemical flows
  - Ultrasonic level sensor on the polymer mixing tank and a pressure sensor on the suction side of the polymer feeder to measure polymer level in the storage drum
  - Ultrasonic level sensors on the alum storage tanks

The plant is typically operated for approximately 12 to 24 hours per day, May through October, and it is the intention to have the facility manned during operation. The plant has storage capacity and the high service pumps can be automatically started and stopped while the plant is unattended to better meet system demands.

The automation project was implemented in a modified design build approach where the engineer designed the automation improvements, procured PLCs, Panels and SCADA system;

programmed and commissioned the plant automation. Field instruments, electrical improvements and installation services were provided under a separate construction contract. The implementation approach involved utility personnel to a great extent and proved to be a successful approach. The active involvement of utility staff also assisted in developing ownership of the new automation systems early in the process.

Project Goals and Benefits

The installation of the base control system was a significant change for the Commission and plant operators in that the existing systems to be replaced were electromechanical and manually operated. The following summarizes several of the goals and the realized/expected benefits:

Goals	Realized/Expected Benefit
Consistency in routine processes	Automation has provided more repeatable operation of the plant vs. manual controls
Workforce changes	With the variation in plant staff skill level the automation helps to reduce training time
Improved water quality consistency	More consistent effluent turbidity
Improved plant reliability	Due to the replacement of older and obsolete equipment.
Enhanced historical data collection	Definitely a benefit achieved over manual data collection. Database driven system allows for the input of manual laboratory data and preparation of compliance reports
Enhanced daily and monthly reports	Report generator provides improved daily reports used to compare the current days' operation to the prior day performance. This assists in developing pertinent operational information for process adjustments. Some of the reports can provide an indication of equipment maintenance issues. Long term historical data and reports can be used for projecting anticipated needs.
Automated control of treatment process	Rapid response to plant upsets through better alarming and automated control responses
Automation hardware standardization	Improve consistency of equipment to reduce design and maintenance costs
Provide for plant expansions	The automation system architecture was planned and implemented with future

	plant expansions and process improvements in mind. The filters and ozone plant improvement projects that followed the control system upgrade were done in a coordinated way with the base automation systems.
Streaming current control	Facilitated the optimization of coagulant dosage thereby reducing chemical costs.

It is worthwhile to note that it has been, and continues to be, a challenge to have staff that can maintain the sophisticated automation systems. To address this issue the Commission has looked to outside firms for instrumentation and control system support.

Project Costs and Estimated Savings

The estimated costs for the Duff automation improvements including engineering and construction was approximately $1.9 M. Although the automation system was not installed with the original goal of unattended plant operation, the level of automation can enable that mode of operation if it is desired in the future.

APPENDIX C
COST DATABASE AND EXAMPLE COST ESTIMATE

COST DATBASE

Samples of the cost database spreadsheets and graphics are included in this Appendix and are included electronically on the attached CD. Several of the detailed cost information spreadsheets were not suitable for printing in this appendix but are included on the Attached CD. The file name is "Appendix C – Cost Database.xls."

Appendix C -- Project Summary
Complete Plant Automation Model Project

			Low		High	
Automation Package	**Qty**	**Unit**	**Unit Cost**	**Extended Cost**	**Unit Cost**	**Extended Cost**
Plant-Wide SCADA System		set	-	-	-	-
Raw Water Pumping Automation Enhancements		set	-	-	-	-
Flocculation & Sedimentation Automation Enhancements (Two Train Set)		set	-	-	-	-
Filter Automation (4 Filter Set)		set	-	-	-	-
Finished Water Pumping Automation Enhancements		set	-	-	-	-
Backwash Water Recovery Automation Enhanacments		set	-	-	-	-
Plant Power Substation Automation Enhancements		set	-	-	-	-
Plant Security Enhnacements		set	-	-	-	-
				-		-
				-		-
				-		-
				-		-
				-		-
				-		-
				-		-
				-		-
				-		-
		Project Total	$	-	to	$ -

Please note that the tool can now be found at
http://www.iwapublishing.com/template.cfm?name=isbn9781843392170

Appendix C -- Automation Package Estimate
Package: Plant-Wide SCADA System

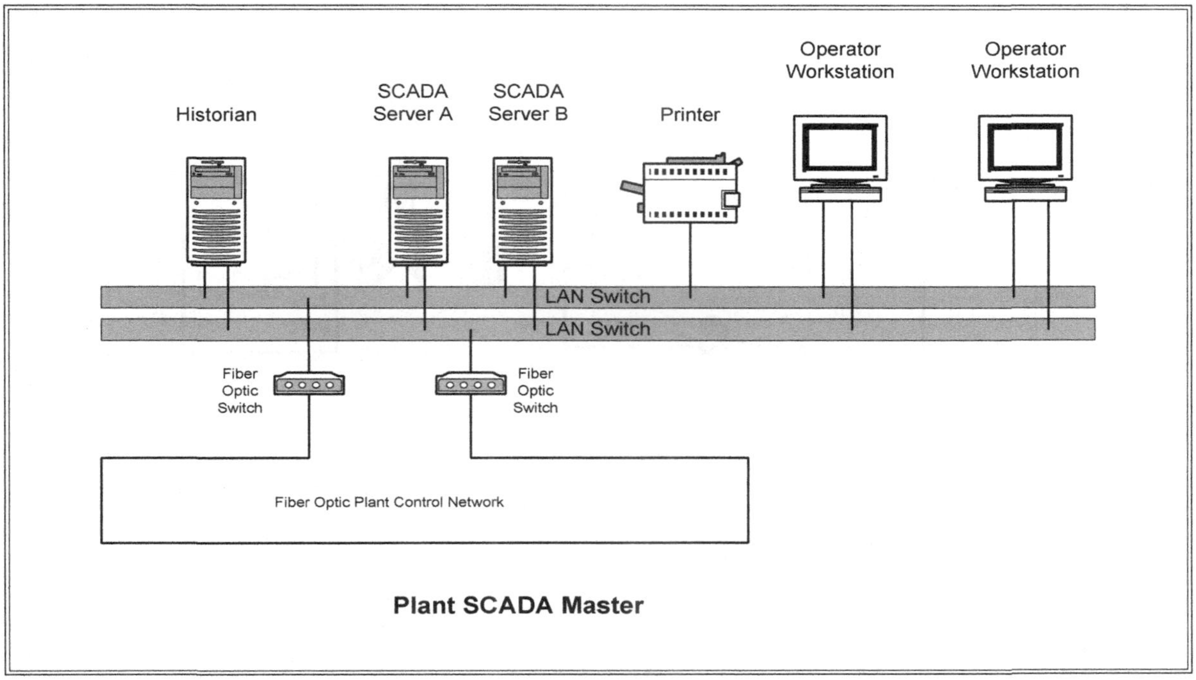

Figure C.1 Typical plant SCADA master schematic

Appendix C -- Automation Package Estimate
Package: Plant-Wide SCADA System

			Low		High	
Work Breakdown	**Qty**	**Unit**	**Unit Cost**	**Extended Cost**	**Unit Cost**	**Extended Cost**
Plant Central SCADA Master with the following:		set	181,697	-	320,294	-
Options:						
Relational History/Reporting System		ea	19,874	-	32,543	-
Simple Report Implementation (each)		ea	728	-	1,305	-
Complex Report Implementation (each)		ea	1,779	-	3,467	-
Remote Access Server		ea	7,150	-	11,385	-
Alarm Notification System		ea	9,195	-	14,329	-
Graphic Display Implementation		ea	462	-	957	-
Plant-Wide Control Network - Fiber Optic:		ea		-		-
Management, Engineering and Administration		lot	8,715	-	14,765	-
Fiber Optic Cable, 4 fiber Multimode (per ft)		ft	9	-	16	-
Raceway, 1 inch Galv. Rigid Steel		ea	14	-	23	-
Fiber Optic Cable Termination Point		ea	426	-	670	-
Process Control Logic Programming:				-		-
PLC Programming/Maintenance Software License		ea	2,727	-	3,679	-
				-		-
			Package Total	$ -	to	$ -

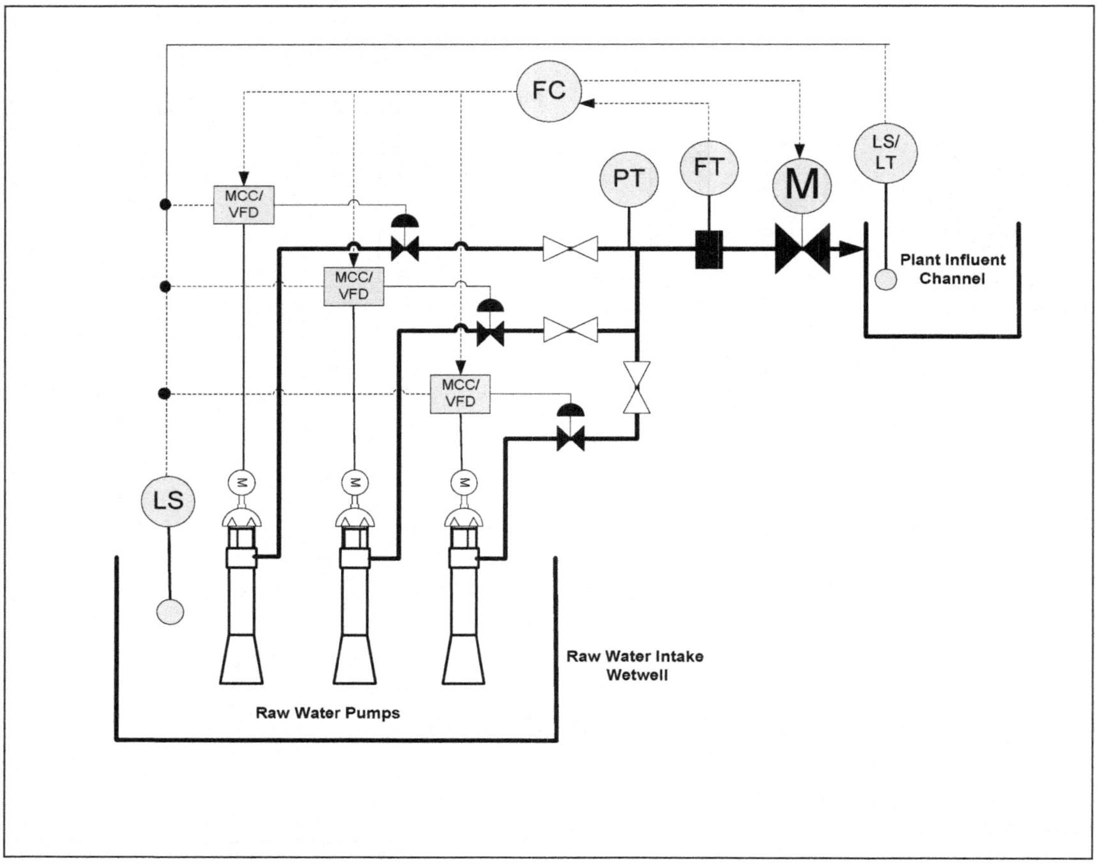

Figure C.2 Raw water pumping automation diagram

Appendix C -- Automation Package Estimate

Package: Raw Water Pumping Automation Enhancements

Work Breakdown	Qty	Unit	Low Unit Cost	Low Extended Cost	High Unit Cost	High Extended Cost
Plant-Wide Control Network - Fiber Optic:						-
Fiber Optic Cable Termination Point		ea	426	-	670	-
Basic Programmable Logic Controller (PLC):						-
Base PLC Unit with Panel Assembly		ea	6,090	-	11,967	-
Discrete Input Point		ea	19	-	40	-
Discrete Output Point		ea	13	-	40	-
Analog Input Point		ea	52	-	100	-
Analog Output Point		ea	147	-	202	-
LCD Panel, 15 inch, Touch Screen		ea	2,761	-	3,768	-
Graphic Display - Develop		ea	398	-	1,045	-
Graphic Display - Apply		ea	331	-	861	-
Process Control Logic Programming:						
Raw Water Pumping Regulation		ea	3,656	-	7,110	-
Process Control & Monitoring						
Magnetic Flowmeter, In-line, 16 to 32 inches		ea	12,409	-	23,551	-
Magnetic Flowmeter, In-line, 8 to 16 inches		ea	9,494	-	14,627	-
Particle Counter		ea	9,133	-	12,509	-
Pressure Transmitter, Relative		ea	2,272	-	3,983	-
Motor Control Interface Wiring & Conduit		ea	408	-	2,666	-
Valve Control Interace Wiring & Conduit		ea	1,583	-	2,922	-
Loop Powered Instrument.Controller Wiring & Conduit		ea	1,511	-	2,823	-
120 VAC Powered Instrument Wiring and Conduit		ea	1,583	-	2,922	-
				-		-
		Package Total	$	-	to $	-

Appendix C -- Automation Package Estimate

Package: Flocculation and Sedimentation Automation Enhancements (Two Train Set)

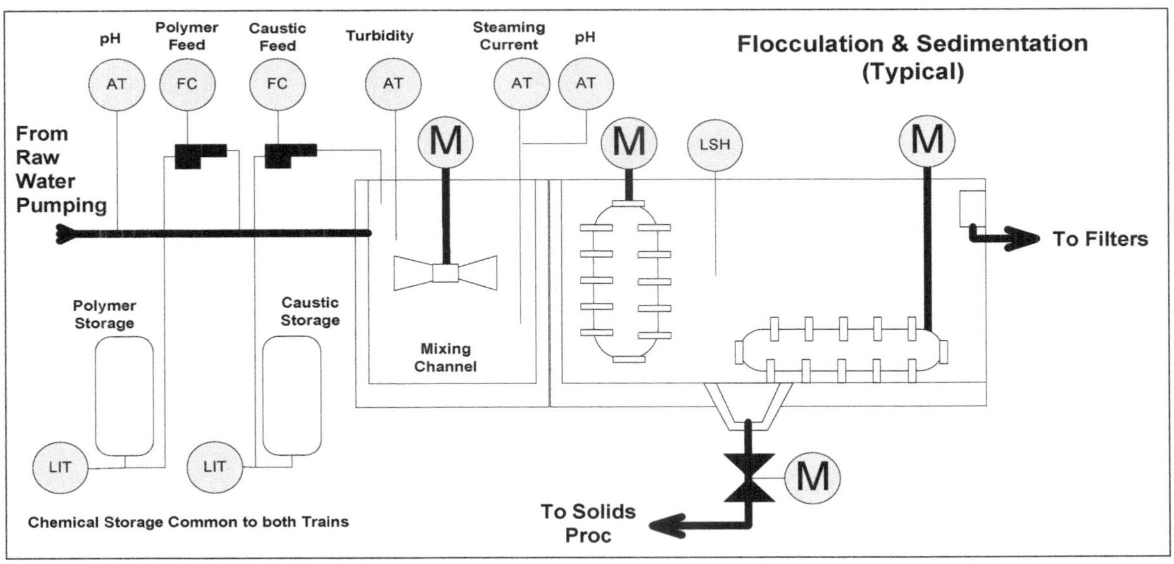

Figure C.3 Flocculation automation diagram

Appendix C -- Automation Package Estimate

Package: Filter Automation Enhancements (4 Filter Set)

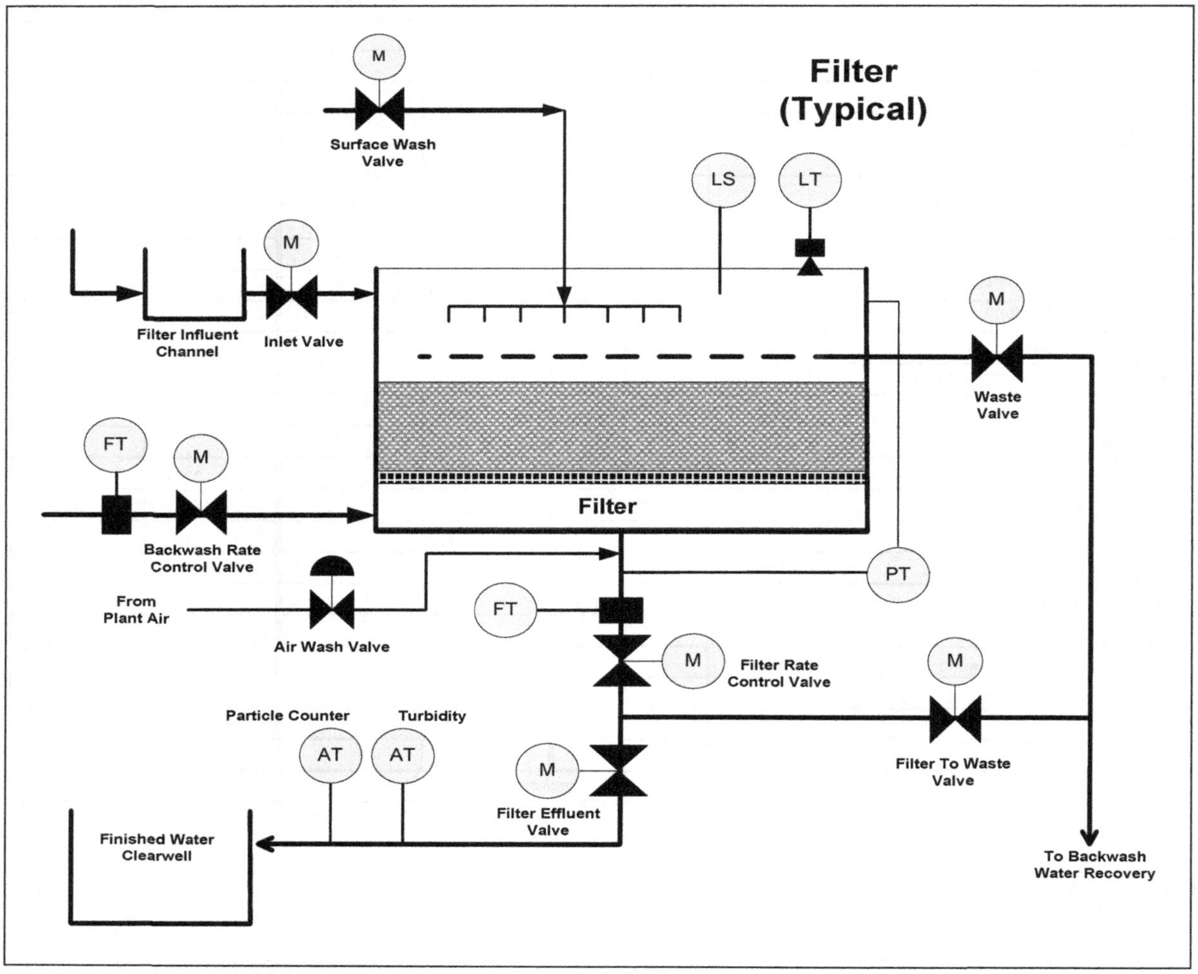

Figure C.4 Filter automation diagram

Appendix C -- Automation Package Estimate

Package: Filter Automation Enhancements (4 Filter Set)

Work Breakdown	Qty	Unit	Low		High	
			Unit Cost	Extended Cost	Unit Cost	Extended Cost
Plant-Wide Control Network - Fiber Optic:						-
Fiber Optic Cable Termination Point		ea	426	-	670	-
Basic Programmable Logic Controller (PLC):						-
Base PLC Unit with Panel Assembly		ea	6,090	-	11,967	-
Discrete Input Point		ea	19	-	40	-
Discrete Output Point		ea	13	-	40	-
Analog Input Point		ea	52	-	100	-
Analog Output Point		ea	147	-	202	-
LCD Panel, 15 inch, Touch Screen		ea	2,761	-	3,768	-
Graphic Display - Develop		ea	398	-	1,045	-
Graphic Display - Apply		ea	331	-	861	-
Process Control Logic Programming:						
Filter Automation - Develop (Runtime and Backwash)		ea	3,746	-	7,282	-
Process Control & Monitoring						
Magnetic Flowmeter, In-line, 4 to 8 inches		ea	6,651	-	11,387	-
Pressure Transmitter, Relative		ea	2,272	-	3,983	-
Turbidity Analyzer, Low Range		ea	3,881	-	6,115	-
Particle Counter		ea	9,133	-	12,509	-
Ultrasonic Level Transmitter		ea	2,625	-	4,523	-
Conductivity Level Switch		ea	1,509	-	2,676	-
Valve Control Interace Wiring & Conduit		ea	1,583	-	2,922	-
Loop Powered Instrument.Controller Wiring & Conduit		ea	1,511	-	2,823	-
24 VDC Powered Instrument Wiring & Conduit		ea	1,583	-	2,922	-
120 VAC Powered Instrument Wiring and Conduit		ea	1,583	-	2,922	-
			Package Total	$ -	to	$ -

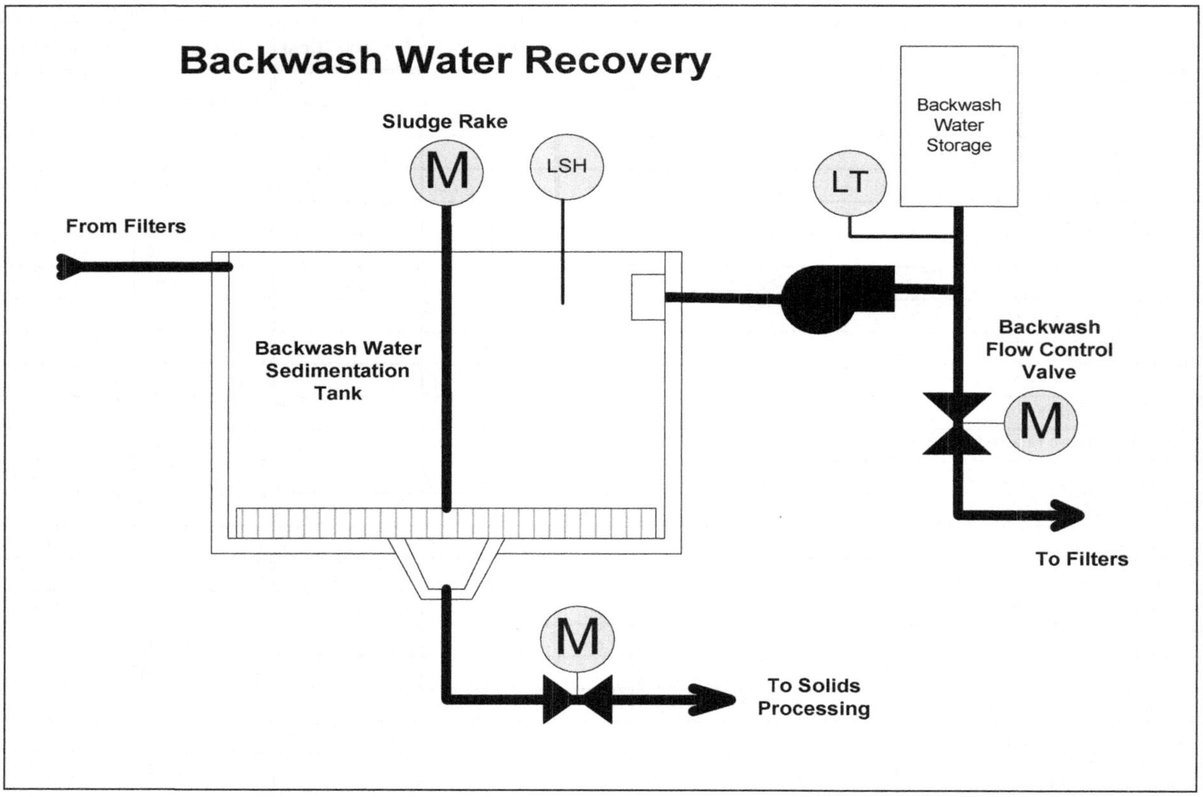

Figure C.5 Backwash Recovery Automation Diagram

Appendix C -- Automation Package Estimate

Package: Backwash Water Recovery

Work Breakdown	Qty	Unit	Low Unit Cost	Low Extended Cost	High Unit Cost	High Extended Cost
Plant-Wide Control Network - Fiber Optic:						-
Fiber Optic Cable Termination Point		ea	426	-	670	-
Basic Programmable Logic Controller (PLC):						-
Base PLC Unit with Panel Assembly		ea	6,090	-	11,967	-
Discrete Input Point		ea	19	-	40	-
Discrete Output Point		ea	13	-	40	-
Analog Input Point		ea	52	-	100	-
Analog Output Point		ea	147	-	202	-
LCD Panel, 15 inch, Touch Screen		ea	2,761	-	3,768	-
Graphic Display - Develop		ea	398	-	1,045	-
Graphic Display - Apply		ea	331	-	861	-
Process Control Logic Programming:						
Backwash Water Recovery		ea	2,777	-	5,930	-
Process Control & Monitoring						
Pressure Transmitter, Relative		ea	2,272	-	3,983	-
Conductivity Level Switch		ea	1,509	-	2,676	-
Motor Control Interface Wiring & Conduit		ea	408	-	883	-
Valve Control Interace Wiring & Conduit		ea	1,583	-	2,922	-
Loop Powered Instrument.Controller Wiring & Conduit		ea	1,511	-	2,823	-
				-		-
				-		-
				-		-
				-		-
				-		-
		Package Total	**$**	**-**	**to**	**$ -**

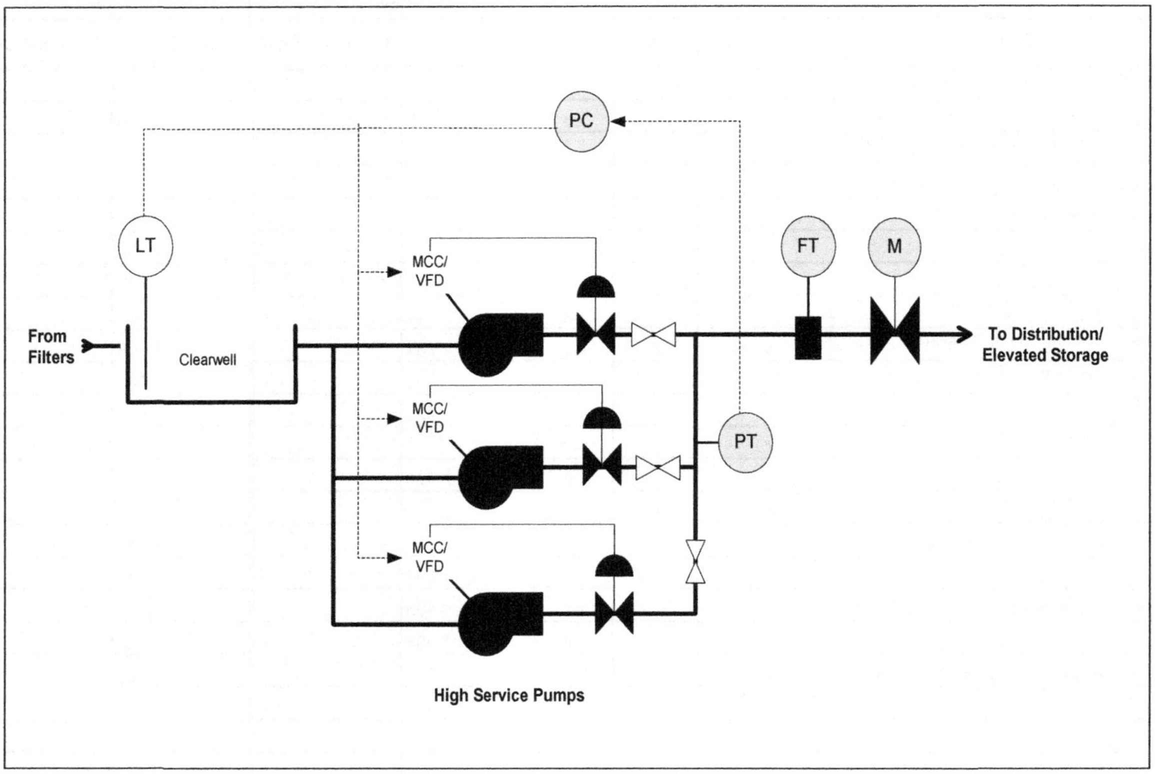

Figure C.6 High service pump automation diagram

Appendix C -- Automation Package Estimate

Package: Finished Water Pumping Automation Enhancements

Work Breakdown	Qty	Unit	Low Unit Cost	Low Extended Cost	High Unit Cost	High Extended Cost
Plant-Wide Control Network - Fiber Optic:						-
Fiber Optic Cable Termination Point		ea	426	-	670	-
###						-
Base PLC Unit with Panel Assembly		ea	6,090	-	11,967	-
Discrete Input Point		ea	19	-	40	-
Discrete Output Point		ea	13	-	40	-
Analog Input Point		ea	52	-	100	-
Analog Output Point		ea	147	-	202	-
LCD Panel, 15 inch, Touch Screen		ea	2,761	-	3,768	-
Graphic Display - Develop		ea	398	-	1,045	-
Graphic Display - Apply		ea	331	-	861	-
Process Control Logic Programming:						
Finished Water Pumping Regulation		ea	3,606	-	7,013	-
Process Control & Monitoring						
Propeller Flowmeter		ea	3,566	-	6,813	-
Pressure Transmitter, Relative		ea	2,272	-	3,983	-
Motor Control Interface Wiring & Conduit		ea	408	-	883	-
Valve Control Interace Wiring & Conduit		ea	1,583	-	2,922	-
Loop Powered Instrument.Controller Wiring & Conduit		ea	1,511	-	2,823	-
Analog Input/Output Loop Wiring, No Conduit		ea	396	-	707	-
				-		-
				-		-
		Package Total	$	-	to $	-

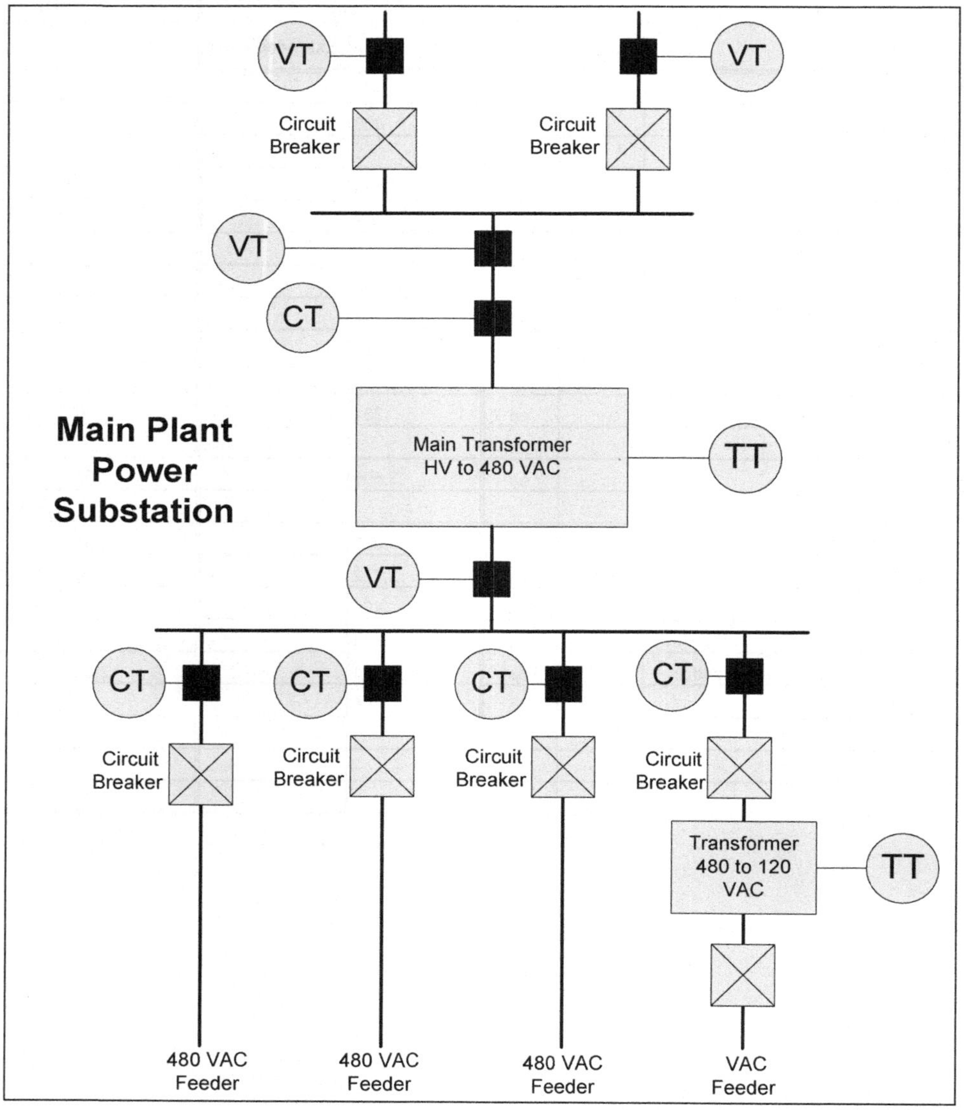

Figure C.7 Power monitoring system diagram

Appendix C -- Automation Package Estimate

Automation Package: Plant Power Monitoring & Control

			Low		High	
Work Breakdown	**Qty**	**Unit**	**Unit Cost**	**Extended Cost**	**Unit Cost**	**Extended Cost**
Plant-Wide Control Network - Fiber Optic:						-
Fiber Optic Cable Termination Point		ea	426	-	670	-
###						-
Base PLC Unit with Panel Assembly		ea	6,090	-	11,967	-
Discrete Input Point		ea	19	-	40	-
Discrete Output Point		ea	13	-	40	-
Analog Input Point		ea	52	-	100	-
Analog Output Point		ea	147	-	202	-
LCD Panel, 15 inch, Touch Screen		ea	2,761	-	3,768	-
Graphic Display - Develop		ea	398	-	1,045	-
Graphic Display - Apply		ea	331	-	861	-
Process Control Logic Programming:						
Substation Power Monitoring		ea	1,234	-	2,553	-
Process Control & Monitoring						
Current/Voltage Transmitter		ea	1,673	-	3,444	-
Temperature Transmitter and RTD Probe		ea	880	-	1,767	-
Loop Powered Instrument.Controller Wiring & Conduit		ea	1,511	-	2,823	-
Analog Input/Output Loop Wiring, No Conduit		ea	396	-	707	-
Discrete Input/Output Wiring, No Conduit		ea	396	-	707	-
				-		-
				-		-
				-		-
			Package Total	**$ -**	**to**	**$ -**

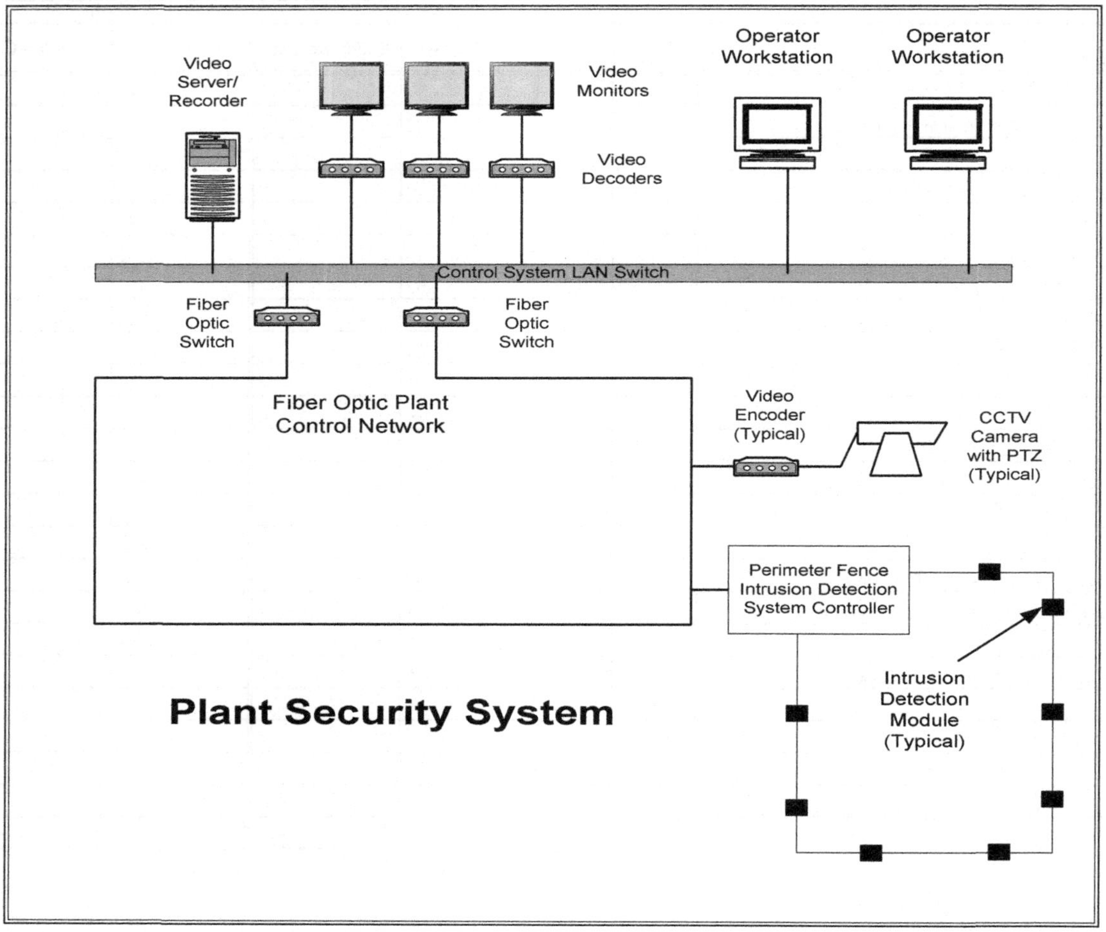

Figure C.8 Security system diagram

Appendix C -- Automation Package Estimate

Plant Security Systems Enhancements

Work Breakdown	Qty	Unit	Low Unit Cost	Low Extended Cost	High Unit Cost	High Extended Cost
Plant-Wide Control Network - Fiber Optic:						-
Fiber Optic Cable Termination Point		ea	426	-	670	-
Intrusion Detection System				-		-
Management, Engineering and Administration		ea	12,056	-	23,098	-
Perimiter Fence System		ea	10,753	-	17,705	-
Perimiter Fence Instrusion Sensing Cable		ft	9	-	16	-
Door/Window Sensor		ea	310	-	623	-
IR Motion Detector		ea	362	-	697	-
CCTV System				-		-
Management, Engineering and Administration			10,291	-	20,201	-
CCTV Color Camera, Fixed, Outdoor			3,776	-	6,623	-
CCTV Color Camera, PTZ, Outdoor			5,089	-	7,800	-
Video Digital Encoder/Decoder			1,477	-	2,553	-
Digital Video Server/Recorder			20,489	-	38,270	-
Digital Video Client Software License			1,050	-	1,766	-
Video Monitor			478	-	1,275	-
				-		-
				-		-
				-		-
				-		-
				-		-
				-		-
				-		-
		Package Total	$	-	to $	-

Appendix E - Cost Database

Labor Rate Schedule

Job Classification	Abbr.	Pay Rate ($/hr.)		Taxes & Fringes (%)		Net Rate ($/hr.)	
		Low	High	Low	High	Low	High
Project Manager	PM	$ 37.00	$ 43.00	37%	42%	$ 51.00	$ 61.00
Project Admin. Assistant	PA	$ 20.00	$ 24.00	37%	42%	$ 27.00	$ 34.00
System/Software Engineer	SE	$ 37.00	$ 43.00	37%	42%	$ 51.00	$ 61.00
Application Engineer	AE	$ 26.00	$ 31.00	37%	42%	$ 36.00	$ 44.00
Field Engineer	FE	$ 29.00	$ 34.00	37%	42%	$ 40.00	$ 48.00
Field Technician	FT	$ 20.00	$ 24.00	37%	42%	$ 27.00	$ 34.00
Production Supervisor	PS	$ 20.00	$ 24.00	37%	42%	$ 27.00	$ 34.00
Production Technician	PT	$ 18.00	$ 21.00	37%	42%	$ 25.00	$ 30.00
Construction Supervisor	CS	$ 48.00	$ 56.00	25%	35%	$ 60.00	$ 76.00
Electrician	EL	$ 44.00	$ 52.00	25%	35%	$ 55.00	$ 70.00
Pipefitter	PF	$ 44.00	$ 52.00	25%	35%	$ 55.00	$ 70.00
Laborer	LB	$ 21.00	$ 25.00	25%	42%	$ 26.00	$ 36.00
Quality Assurance Inspector	QA	$ 26.00	$ 31.00	37%	42%	$ 36.00	$ 44.00

COST ESTIMATE EXAMPLE

The following spreadsheet printouts are for the budgetary construction cost estimate for the Roxborough example provided n Appendix A. Electronic copies of this example spreadsheets are included in the attached CD, file name Roxborough Implementation Estimate.xls.

Appendix C -- Project Summary
Complete Plant Automation Model Project
Roxborough Business Case Example

Automation Package	Qty	Unit	Low Unit Cost	Low Extended Cost	High Unit Cost	High Extended Cost
Plant-Wide SCADA System		set	-	-	-	-
Raw Water Pumping Automation Enhancements	1	set	12,640	12,640	20,472	20,472
Flocculation & Sedimentation Automation Enhancements (Two Train Set)	1	set	19,246	19,246	32,073	32,073
Filter Automation (4 Filter Set)		set	-	-	-	-
Finished Water Pumping Automation Enhancements		set	-	-	-	-
Backwash Water Recovery Automation Enhanacments		set	-	-	-	-
Plant Power Substation Automation Enhancements		set	-	-	-	-
Plant Security Enhnacements		set	-	-	-	-
				-		-
				-		-
				-		-
				-		-
				-		-
				-		-
				-		-
				-		-
				-		-
		Project Total	**$**	**31,886**	**to**	**$ 52,545**

MIDPOINT **$ 42,215**

Appendix C -- Automation Package Estimate

Package: Raw Water Pumping Automation Enhancements

Work Breakdown	Qty	Unit	Low Unit Cost	Low Extended Cost	High Unit Cost	High Extended Cost
Plant-Wide Control Network - Fiber Optic:						-
Fiber Optic Cable Termination Point		ea	426	-	670	-
Basic Programmable Logic Controller (PLC):						-
Base PLC Unit with Panel Assembly		ea	6,090	-	11,967	-
Discrete Input Point		ea	19	-	40	-
Discrete Output Point		ea	13	-	40	-
Analog Input Point	1	ea	52	52	100	100
Analog Output Point		ea	147	-	202	-
LCD Panel, 15 inch, Touch Screen		ea	2,761	-	3,768	-
Graphic Display - Develop		ea	398	-	1,045	-
Graphic Display - Apply		ea	331	-	861	-
Process Control Logic Programming:						
Raw Water Pumping Regulation		ea	3,656	-	7,110	-
Process Control & Monitoring						
Magnetic Flowmeter, In-line, 16 to 32 inches	0	ea	12,409	-	23,551	-
Magnetic Flowmeter, In-line, 8 to 16 inches	1	ea	9,494	9,494	14,627	14,627
Particle Counter		ea	9,133	-	12,509	-
Pressure Transmitter, Relative		ea	2,272	-	3,983	-
Motor Control Interface Wiring & Conduit		ea	408	-	2,666	-
Valve Control Interace Wiring & Conduit		ea	1,583	-	2,922	-
Loop Powered Instrument.Controller Wiring & Conduit	1	ea	1,511	1,511	2,823	2,823
120 VAC Powered Instrument Wiring and Conduit	1	ea	1,583	1,583	2,922	2,922
				-		-
		Package Total	$	12,640	to	$ 20,472

Package: Flocculation and Sedimentation Automation Enhancements (Two Train Set)

			Low		High	
Work Breakdown	**Qty**	**Unit**	**Unit Cost**	**Extended Cost**	**Unit Cost**	**Extended Cost**
Plant-Wide Control Network - Fiber Optic:						-
Fiber Optic Cable Termination Point		ea	426	-	670	-
						-
Base PLC Unit with Panel Assembly		ea	6,090	-	11,967	-
Discrete Input Point	0	ea	19	-	40	-
Discrete Output Point		ea	13	-	40	-
Analog Input Point	1	ea	52	52	100	100
Analog Output Point	1	ea	147	147	202	202
LCD Panel, 15 inch, Touch Screen		ea	2,761	-	3,768	-
Graphic Display - Develop	2	ea	398	796	1,045	2,090
Graphic Display - Apply	2	ea	331	661	861	1,722
Process Control Logic Programming:						
Flocculation Automation - Develop	1	ea	3,746	3,746	7,282	7,282
Flocculation Automation - Apply	1	ea	3,004	3,004	5,799	5,799
Process Control & Monitoring						
pH Analyzer		ea	2,830	-	4,694	-
Turbidity Analyzer, High Range		ea	6,507	-	9,667	-
Streaming Current Meter	1	ea	6,507	6,507	9,133	9,133
Pressure Transmitter, Relative		ea	2,272	-	3,983	-
Capacitance Level Switch		ea	1,509	-	2,676	-
Motor Control Interface Wiring & Conduit		ea	408	-	883	-
Valve Control Interace Wiring & Conduit		ea	1,583	-	2,922	-
Loop Powered Instrument.Controller Wiring & Conduit	1	ea	1,511	1,511	2,823	2,823
120 VAC Powered Instrument Wiring and Conduit	1	ea	2,823	2,823	2,922	2,922
		Package Total	**$**	**19,246**	**to**	**$ 32,073**

APPENDIX D
LITERATURE RESEARCH AND REVIEW

Literature Research

A literature search and review of water utility and non-water utility automation was conducted for the AwwaRF Project 3019 "Costs and Benefits of Complete Water Treatment Plant Automation". This chapter identifies the objectives of the literature review, discusses the methodology and approach and summarizes the relevant findings. In addition to the findings summarized in this chapter, a detailed bibliography is included at the end of the report. The bibliography is organized by industry and includes a citation and brief overview indicating relevance to the research project.

Objectives

The literature review was conducted to gain an understanding of the issues related to automation in the water industry with a focus on analysis of the costs, benefits and risks associated with complete water plant automation. In addition to the water utility literature review, a non-water utility literature review was conducted to compile information from other industries and identify experiences and lessons learned that can be useful for the water industry. Primary objectives of the literature research included:

- Review of documents describing the level of automation available to water and non-water utilities including case studies and success stories
- Identification of current methodologies and approaches being taken to automate treatment plants
- Review of the requirements for automation for all types of treatment plant processes being utilized in the water treatment industry
- Identification of current methodologies and approaches being taken to automate other types of plants and facilities
- Review of current and pending regulations that impact the level of automation that utilities can utilize
- Review of risk assessment methodologies performed in other industries
- Review of previous cost/benefit analysis work done within the Water Utility community and non-water utilities.

Non-water utilities researched included: wastewater treatment, industrial waste processing, hydroelectric power, coal and oil fired power plants, petrochemical industry and business information technology.

Approach

A variety of resources including library indexes, internet search engines, industry journals and databases, other research literature reviews and bibliographies from relevant works were consulted to conduct the literature research. Searches were performed using keywords derived from various databases. The keywords are listed below to aid the user in searching for additional citations from the literature.

- Water Treatment
- Water Utilities
- Water Distribution
- Wastewater Treatment
- Industrial Waste
- Power Plants
- Hydro Electric Power
- Automation
- Process Control Systems
- Instrumentation and Control Systems
- SCADA
- Costs and Benefits
- Cost Data
- Economic Analysis
- Risks and Barriers
- Risk Mitigation
- Case Studies

Search results were analyzed and relevant articles were obtained and reviewed to establish a body of literature used as a basis for the research. Important sources for the water utility research included Journal AWWA, various AWWA Conference Proceedings, AWWA Research Foundation research reports, Water Science and Technology, Water Engineering and Management, and EPA reports. Key sources for the non-water utilities included WEFTEC Conference Proceedings, Water Environment and Technology, IEEE Conference Proceedings, and various industry journals. A bibliography was created for the body of literature which was organized by industry and included the citation and a brief overview of the document and relevance to the research.

LITERATURE REVIEW AND BIBLIOGRAPHY

This bibliography is a compilation of the literature review conducted for the AwwaRF Project 3019 "Costs and Benefits of Complete Water Treatment Plant Automation" and includes literature citations, brief reviews and a discussion of the significance to the research project. References are sorted alphabetically by author within major subdivisions corresponding to the relevant industry.

Water Utility Industry

AWWA (American Water Works Association). 2001. *Instrumentation and Control,* AWWA Manual M2, Third Edition. Denver, Colo.: AWWA.

AWWA (American Water Works Association). 2000. *Operational Control of Coagulation and Filtration Processes*, AWWA Manual M37, Second Edition. Denver, Colo.: AWWA.

Describes in detail methods used to optimize coagulation and filter processes. The use of streaming current detectors and particle counters for process optimization and problem detection are covered in detail. The use of Pilot plants as a tool for early detection of upset and failure situations is discussed.

AWWA (American Water Works Association). 2005. *Water Treatment Plant Design, Fourth Edition.* New York: McGraw-Hill.

Provides in-depth design discussions related to each treatment plant process and includes a chapter on Process Instrumentation and Controls.

Barnes, M., C. Brophy and R.J. Daly. 1997. Automating the Lake DeForest Treatment Plant: A Step-Wise Approach to Unattended Operations. In *Proc. 1997 AWWA Computer Conference, April 13-16, 1997, Austin, Texas.* Denver, Colo.: AWWA.

The paper discusses the planning and design of an automated process control system for the 20 mgd DeForest Treatment Plant operated by United Water New York. The project was completed in two phases with the ultimate goal to permit extended periods of unattended operation – nights, weekends and several hours during the day which significantly reduces manpower costs.

Baumeister, R.A. 1996. Treatment Plant Automation: Implications for Treatment Optimization. In *Proc. 1996 AWWA Annual Conference and Exposition, June 23-27, 1996, Toronto, Ontario, Canada.* Denver, Colo: AWWA.

This paper discusses how water treatment automation and treatment optimization relate and discusses unattended operation of surface water treatment plants.

Bevan, D., C. Cox, and A. Adgar. 1998. Implementation Issues when Installing Control and Condition Monitoring at Water Treatment Works. In *IEE Colloquium on Industrial Automation and Control: Distributed Control for Automation (Digest No. 1998/297), March 4, 1998, London, UK.* IEE.

This article provides an overview of a basic three-stage water treatment process and classifies four levels of automation that may be used to monitor and control the process. The author states that there is wider variability in the level of automation for the water industry than in other industries due to complexity of process measurement and variability of source water. The author further states that generally the level of instrumentation at a WTP is often the limiting factor in the level of automation and identifies problems that can arise when converting monitoring-only plants to a high level of automation.

Brooks, R.L. 1992. Operational Cost Savings from SCADA Systems – Fact or Fiction? In *Proc. of 1992 AWWA Computer Conference, April 12-15, 1992, Nashville, Tennessee.* Denver, Colo.: AWWA.

This paper examines the results of one water utility's reevaluation of a multimillion dollar SCADA project after the system was operations for one year.

Brown, J. 1989. Control System Conserves Staff Time, Saves Maintenance Costs. *Waterworld News,* 5(3):32.

Article describes the Wolf Creek Highway Water District SCADA System which allows the District to operate its facilities unattended on nights, weekends and holidays resulting in significant cost savings.

Cascos, G. and T. DeLaura. 2003. Improving Energy Efficiency Using a Department-Wide SCADA System. In *Proc. AWWA Annual Conference and Exposition, June 15-19, 2003, Anaheim, CA.* Denver, Colo: AWWA.

This paper discusses specific control strategies implemented in a new SCADA System for the Detroit Water and Sewerage Department, that were designed to lower energy and other operating costs. The paper includes example cost savings that can be realized from peak shaving with the use of emergency backup generators.

Copithorn, R.R. and J.K. Warrick. 1989. Automation of a Water Treatment Plant for Unattended Operation and Shutdown. In *Proc. AWWA Computer Specialty Conference, April 2-4, 1989, Denver, Colorado.* Denver, Colo: AWWA.

This paper presents the automation for a water treatment plant in upstate New York that allowed the plant to be operated unattended and shut down automatically when the main finished water storage tank is filled. The paper describes the automation components, operational strategies, costs and benefits and future plans for automation.

Covelli, D. 2003. Nation's Largest AMR System Comes Online in DC. *Water & Wastes Digest,* 43(7).

This article summarizes a large automated meter reading (AMR) project implemented by the District of Columbia Water and Sewer Authority (WASA) to replace aging, inaccurate water meters. WASA reported that the revenue improvement provided by the AMR will cover the cost of the meter replacement program.

Critchley, R.F., E.O. Smith and P. Pettit. 1990. Automatic Coagulation Control at Water Treatment Plants in the North-West Region of England. *Journal of the Institution of Water and Environmental Management,* 4(6):535-543.

This article discusses evaluations performed by North West Water for several automatic coagulation control systems covering a range of raw waters and treatment facilities. It is shown that under appropriate conditions, streaming current type systems are effective for automated control of coagulant dose. The benefits of automated control are discussed include savings in coagulant usage.

Dentel, S.K., and K.M. Kingery. 1988. *An Evaluation of Streaming Current Detectors.* Denver, Colo.: AwwaRF and AWWA.

This paper reviews the effect streaming current analyzers have on coagulation dosage control.

Ehlen, Dana J., Mark Maxwell, and John Schmitz. 1992. Custom-Blended Coagulant Reduces Costs, Improves Water Quality. *Water Engineering & Management,* 139(12):17.

This paper describes the City of Aurora, Colorado's experiences in developing a custom blending technique for alum and cationic polymer which resulted in dramatically reduced chemical costs. The system was implemented at a 60 MGD plant and a 70 MGD plant. Coagulant dosing was reduced by 63%. Finished water turbidity decreased by 47 percent. Filter run times have tripled due to a dramatic reduction of aluminum hydroxide loading. This resulted in a net 2.5% increase in plant throughput. Annual Cost savings included $68,000. for coagulant chemicals (decreased from $8.46 to $1.88 per million gallons), $31,200. due to reductions in backwash water volume (from 4.5% to 2% of treated water production), and $2,900. for reduced solids handling. The paper does not discuss automation or unattended operation.

Emanuel, R.C., and B. Beaudet. 2003. Designing Automation Facilities. In *Proc. AWWA Annual Conference and Exposition, June 15-19, 2003, Anaheim, CA.* Denver, Colo: AWWA.

This paper provides an overview of primary considerations for defining criteria, planning, implementing and sustaining a successful automation project in the water industry. Benefits and risks of varying levels of automation are discussed. The author asserts that selective application of automation along with operational and long term planning strategies could allow utilities to achieve significant cost saving efficiencies that would offset the cost of implementing the automation.

EPRI (Electric Power Research Institute). 1994. *Energy Audit Manual for Water/Wastewater Facilities: A Guide for Electric Utilities to Understanding Specific Unit Processes and Their Energy/Demand Relationships at Water and Wastewater Plants*. CR-104300. Palo Alto, Calif.: EPRI.

EPRI (Electric Power Research Institute). 1996. *Water and Wastewater Industries: Characteristics and Energy Management Opportunities*. CR-106941. Palo Alto, Calif.: EPRI.

EPRI (Electric Power Research Institute). 1997. *Quality Energy Efficiency Retrofits for Water Systems*. CR-107838. Palo Alto, Calif.: EPRI.

EPRI (Electric Power Research Institute). 2001. *Summary Report for California Energy Commission Energy Efficiency Studies*. Palo Alto, Calif.: EPRI.

Freeman, I.C., and S.J. Prutz. 2004. Best Practices for the Reduction of SCADA System Total Lifecycle Cost. In *Proc. AWWA Annual Conference and Exposition, June 13-17, 2004, Orlando, FL*. Denver, Colo: AWWA.

This paper summarizes a study which identified best practices for reduction of total lifecycle costs for SCADA systems for water utilities. Six phases of total life cycle costs are defined and cost drivers are identified. The study employed a systems engineering technique known as Quality Function Deployment (QFD) to analyze a two-part survey and determine best practices. The 25 best practices for total life cycle costs reduction are identified, while the top 10 are discussed in detail.

Frey, M. and L. Sullivan. 2004. *Practical Application of Online Monitoring*. Denver, Colo.:AwwaRF and AWWA.

The report evaluates how utilities are using and maintaining their online equipment and provides guidelines for utilities to consider for their online monitoring program. The study included a survey of 264 utilities.

Gotoh, K., J.K. Jacobs, S. Hosoda, and R.L. Gertsberger, eds. 1993. *Instrumentation and Computer Integration of Water Utility Operations*. Denver, Colo.: AwwaRF.

This book, funded in a joint effort by AwwaRF and JWWA, is a comprehensive review of computer-based monitoring, control and management of water utility operations. Although published in 1993, many of the principles and concepts are still very relevant and serve as a basis of understanding for automation in water utilities. Topics include application of technology and principles, description of process control strategies, discussion of CIS, AM/FM/GIS, LIMMS, MMS, leakage control systems and emergency response and discussion of water utility challenges. Future trends and research needs are identified for each of the areas.

Graney, G.D. 2000. Optimizing the "True Cost of Ownership" for SCADA and Automation Solutions Supporting Water Treatment Plants. In *Proc. 2000 AWWA Information Management & Technology Conference, April 16-19, 2000, Seattle, Washington.* Denver, Colo.: AWWA.

This paper presents some of the issues that must be considered in terms of planning, acquiring or upgrading a SCADA solution for a water treatment plant. The author defines the true cost of ownership (TCO) and that optimized TCO does not necessarily mean lowest cost. Also TCO factors for a small application are considerably different that the factors involved in automation of a large WTP that includes construction elements and a multi-year life cycle.

Great Lakes – Upper Mississippi River Board of State and Provincial Public Health and Environmental Managers (Ten State Standards). 2003. *Recommended Standards for Water Works, Policy Statement on Automated/Unattended Operation of Surface water Treatment Plants.* Albany, NY: Great Lakes – Upper Mississippi River Board of State and Provincial Public Health and Environmental Managers.

The policy statement recommends that an engineering report be developed prior to the design of automation systems for water treatment plants utilizing automated/unattended operations. The policy further identifies the information and criteria that should be included in an engineering report.

Hardison, J. and R. Martin. 2000. Plant SCADA System Enables Optimization Initiative. In *Proc. 2000 AWWA Annual Conference and Exposition, June 11-15, 2000, Denver, Colorado.* Denver, Colo.: AWWA.

Paper discusses the experiences of Colorado Springs Utilities as their conventional PLC–based process control system evolved into a fully-integrated facility management system. The human-machine interface functions as the common operating environment for plant operations, maintenance and process management.

Hargesheimer, E., O. Conio and J. Popovicova. 2002. *Online Monitoring for Drinking Water Utilities.* Denver, Colo.: AWWARF, AWWA and CRS PROAQUA.

Comprehensive reference book dealing specifically with drinking water online monitoring for water utilities.

Hinthorn, R., F. Moshavegh, L. Wilson, S. Yadav. 2003. A Vision for Real-time Monitoring and Modeling of Water Quality in Water Distribution Systems. In *Proc. of AWWA Information Management Technology Conference, April 27-30, 2003, Santa Clara, CA.* Denver, Colo.: AWWA.

This paper presents a vision for real-time monitoring and modeling of water distribution system hydraulics and water quality. The paper defines real-time monitoring and modeling, identifies the components and effort needed to implement such a system, and provides an overview of the City of San Diego Water Department (CSDWD) integrated system. The paper also identifies the benefits and value that can be realized and the challenges that accompany these systems.

Hoffman, M.R. and L.K. Reynolds. 1994. The Role of Control Systems in the Management of a Modern Water Utility. In *Proc. of IEEE Conference on Control Applications, August 24-26, 1994, Glasgow, UK.* IEEE.

This paper highlights a broad range of process control applications in the water and wastewater industry using the Thames Water Utilities (TWU) as a case study. The paper also discusses TWU's approach to strategic planning and cost/benefit analysis for control applications and improvements. TWU has accelerated "demanning" of facilities and all new plants and processes are designed for unattended operations.

Huntington, R. 1998. Twenty Years Development of ICA in a Water Utility. *Water Science and Technology,* 37(12):27-34.

This paper describes the long term planning and development of automation improvements in a large water and wastewater utility, the Wessex Water Authority, and identifies benefits realized from the improvements. Although details of the cost benefit analysis are not provided, the author indicates that many of the improvement projects provide pay back in five years. Benefits of improved water qualify were quantified by increases in percentage of compliance. Treated water went from 85% compliance to 99.98% and wastewater quality went from less than 50% to 98.6% compliance.

Jentgen, L., S. Conrad, and T. Lee. 2005. Optimizing System Operations. *Jour. AWWA,* 97(8): 58-66.

This paper describes an operations optimization model (or energy management system) used by two large water utilities to reduce operating costs and defer capital expenditures. This is an example of an advanced automation application that utilizes SCADA data to monitor and forecast operational needs and improve planning. Case studies sited saved $1.1 million in energy and water supply costs in the first seven months and $1.4 million in deferred capital costs.

Jacobs, J., T. Kerestes and W.F. Riddle. 2003. *Best Practices for Energy Management.* Denver, Colo.: AwwaRF.

This report provides a methodology for energy management at water utilities.

Johnson, D. 2004. Beyond Process Control. *Water and Wastewater Products, 4(6).*

This article discusses the drivers and benefits of integrating SCADA and plant controls with asset management, process optimization and enterprise resource planning management.

Junnier, R.J. 1997. Developing Cost/Benefits for the Implementation of Computer Technology ay Water & Wastewater Utilities. In *Proceedings of Computer Technologies for the Competitive Utility, Water Environment Federation, July 15-18, 1997, Philadelphia, Pennsylvania.* Alexandria, Va.: WEF.

This paper describes a methodology approach to developing a cost/benefit analysis for the implementation of computer technology for water and wastewater utilities. Areas investigated included computerized maintenance management systems, inventory management systems, geographic information systems, SCADA and automated O&M manuals.

Kingham, T.J. and T.M. Hoggart. 1995. Chlorination Control in a Large Water Treatment Works. In *IEE Colloquium on Application of Advanced PLC Systems with Specific Experiences from Water Treatment (Digest No. 1995/112), June 29, 1995, London, UK.* IEE.

This paper provides detailed control strategies for control of water chlorination that is being implemented for a large U.K water treatment facility.

Lauer, W.C. 2005. Automation Supports Unattended Operation. *Opflow,* 31(2):7.

This article provides insight into data found in the US Environmental Protection Agency 2000 Community Water System Survey related to unattended operation of water treatment plants.

Maxwell, S. 2005. An Overview of Trends in the Drinking Water and Wastewater Treatment Markets. *Jour. AWWA*, 97(9): 4-5.

McIntyre, J. 2003. Water Treatment Regulation in the United States and Effects on the Global Innovation Process. Working Paper. Georgia Institute of Technology, Atlanta, Ga.

This article provides history and background information on water treatment regulation in the United States. Several innovative process technologies which resulted from enforcement of water treatment regulations are discussed. One of the innovations discussed is a mobile monitoring lab which uses real-time data to diagnose problems and analyze plant performance.

Means III, E.G., L. Ospina, and R. Patrick. 2005. Ten Primary Trends and Their Implications for Water Utilities. *Jour. AWWA,* 97(7): 64-77.

This article summarizes an AwwaRF project which identifies and characterizes future trends of importance to the public water supply community and discusses strategies to address the trends. One of the top ten trends identified in the article is technological advances. The study predicts that automation will grow in importance and utilities will explore having minimally or unattended operations. The study recommends that utilities strategically apply technology in all areas of utility business and operations and develop a plan to utilize technology to reduce costs associated with labor, chemicals and energy usage.

Means III, E.G., J. Bernosky, and R. Patrick. 2006. Technology Trends and Their Implications For Water Utilities. *Jour. AWWA*, 98(1): 60-71.

National Drinking Water Clearinghouse. 1997. Package Plants. *Tech Brief*, November 1997, 6:1-4.

This paper describes the various types of package water treatment plants available, limitations, advantages and design and operation considerations. The paper includes a discussion of the automated controls which allow unattended operation of the package plant.

Ohto, T. 1998. Controls, Computers and Communications: Fusion in Instrumentation, Control and Automation of Water and Wastewater Systems in Japan. *Water Science and Technology.* 37(12):15-19.

This article describes the current status of water and wastewater, instrumentation, control and automation in Japan and new developments which the author describes as '3C Fusion' (i.e. fusion of controls, computing and communications.) By expanding the Wastewater Work's MAN with optical fiber in sewers, operators will be able to monitor and control remote facilities using realtime video and sound.

Opincar, V. 2003. Automated Surveillance for Water Utilities. *Jour. AWWA,* 95(9):48-51.

Video surveillance is a common recommendation made in vulnerability assessments in the past few years. However, this requires assigning personnel to monitor the video. This can decrease the effectiveness of water utility operations if this task is added to the existing duties of operations staff. This article discusses a specific technology used to automate surveillance for water utilities. The system utilizes video-processing algorithms that identify anomalous activity within regions of interest on a video image. The significance of the article is that is reduces the reliance on 24/7 guard force and more effectively utilizes existing staff to monitor video surveillance. The author asserts that a 10 year cost for a 24/7 guard force is in excess of $13 million.

Preble, C. and T. Valorose. 1995. Computer at Water Plant Runs Second Plant by Remote Control. *Water Engineering & Management,* 142(5):34-38.

This article discusses the history and control philosophy changes that were made by a small water utility in Wilmington, MA which led to control system capabilities allowing unattended operations at a second water treatment plant. The article describes the system architecture, hardware, software and communication systems installed to support remote monitoring and control.

Rau, M. 1990. Building Momentum for Automatic Meter Reading. *Water Engineering & Management,* 137(3):32-33.

This article discusses the approach that Waukesha Water Utility took to implement an automatic meter reading system, focusing on breaking down barriers to the implementation. The utility involved customers early in the process, addressing their concerns and obtaining feedback and buy-in to support the full-scale implementation.

Renner, R., B. Hegg, and J. Bender. 1990. Optimizing Water Treatment Plant Performance With the Composite Correction Program. EPA/625/8-90/017. Washington D.C. USEPA.

A Technology Transfer Summary Report which summarizes the results of an ongoing project to evaluate effectiveness of the Composite Correction Program (CCP) approach to improving the performance of drinking water treatment facilities. The CCP approach is described and the results of evaluating it at 13 drinking water plants is summarized. The case studies focus on the potential of CCP to improve the performance of small drinking water systems in meeting the turbidity removal requirements of the Surface Water Treatment Rule (SWTR).

Rice, M.D., R. Edmonson, and L.R. Smith. 2005. Single Source Responsibility = Seamless SCADA Expansion and Integration. In *Proc. of AWWA Annual Conference and Exposition, June 12-16, 2005, San Francisco, CA.* Denver, Colo.: AWWA.

This paper describes a mitigation strategy used by Dallas Water Utilities (DWU) to lower the risk of a potentially unsatisfactory SCADA implementation that does not meet expected performance requirements. Anticipating a phased implementation with multiple contracts, DWU decided to partner with a single entity to provide the planning, design, application programming, startup, testing and training for the new SCADA system.

Schlenger, D.L., and W.F. Riddle. 1993. Can Computers Replace People in Water Treatment Operations? Guidelines for Staffing Adjustments Through Automation. *In Proc. 1993 AWWA Annual Conference, June 6-10, 1993, San Antonio, Texas.* Denver, Colo.: AWWA.

This study investigated automation management strategies for water treatment facilities. The study included a survey of water quality regulators regarding automation in the stats, a discussion of benefits and costs of automation, an automation decision model, maintenance and human factors.

Schlenger, D.L., W.F. Riddle, B.K. Luck, and M.H. Winter. 1996. *Automation Management Strategies for Water Treatment Facilities.* Denver, Colo.: AwwaRF and AWWA.

This research project identifies current (1996) water industry automation practices, investigates the factors affecting automation decisions and presents a model for making automation decisions. The study recommends a present value economic analysis using control system life-cycle costs. The study recognizes that intangible benefits should be considered in the cost-benefit analysis and recommends further research to define cost data, establish estimating guidelines and test the economic principles. The study also describes specific control strategies for water treatment applications, identifies human factors and attitudes that affect automation decisions, and includes case studies and interview information. This research project provides a basis of understanding and is a precursor to Project 3019 – Costs and Benefits of Complete Water Treatment Plant Automation.

Schoedinger, S. and D. Diffee. 1999. Water/Wastewater Treatment Plant Automation Strategies for Competitiveness and Efficiency. In *Proc. Florida Section AWWA Annual Conference, November 29 – December 2, 1999, Orlando, FL.* Denver, Colo.: AWWA.

This paper identifies a number of significant opportunities to capture both short-term and long-term savings through enhanced automation of processes for the Miami-Dade Water and Sewer Department (MDWASD) water and wastewater treatment plant. The paper identifies O&M cost, chemical usage and energy savings generated through automation efficiency improvements. Longer term savings are identified by installing distributed control systems at each of MDWASD's plants.

Shariff, R., R. Welz, S. Stanley, W. Stachowski, and R. Corscadden. 2001a. Remote Monitoring and Operation of Isolated Facilities in Cold Regions. In *Proceedings of AWWA Annual Conference and Exposition, June 17-20, 2001, Washington D.C.* Denver, Colo.: AWWA.

A pilot study was performed to test on-line analyzers and RTUs connected to a SCADA system for remote, unattended operations. The study included a low cost approach to a wide area SCADA system and centralized control center for small isolated water plants.

Shariff, R., R. Welz, R., S. Stanley, W. Stachowski, and R. Corscadden. 2001b. Automation and Unattended Operation of Large Water Plants. In *Proceedings of AWWA Information Management Technology Conference, April 8-11, 2001, Atlanta, Georgia.* Denver, Colo.: AWWA.

This paper describes in detail the analysis and approach that EPCOR utilized to implement unattended operation at the EL Smith Water Treatment Plant. This paper has significant relevance to this research project.

Shariff, R., A. Cudrak, W. Stachowski, R. Welz, and S. Stanley. 2002. Centralized Remote Operation of Multiple Water and Wastewater Plants. In *Proc. of AWWA Information Management Technology Conference, April 14-17, 2002, Kansas City, MO.* Denver, Colo.: AWWA.

This paper discusses the methods and technologies used in developing and implementing a scalable SCADA system for operating multiple, remotely located water and wastewater treatment plants from a centralized location. The case study is based on EPCOR which operates several small water and wastewater treatment plants in Alberta and British Columbia, Canada. The solution incorporated the use of virtual private networks (VPNs) via the internet, a wide area SCADA system and increased use of on-line analyzers for water quality monitoring . The paper identified criteria, strategies to improve reliability, and benefits of centralized control.

Shariff, R., A. Cudrak, E. Saumer, and S. Stanley. 2003. Improve Performance and Decision-Making Through Greater Utilization of Plant Data. In *Proc. of AWWA Information Management Technology Conference, April 27-30, 2003, Santa Clara, CA.* Denver, Colo.: AWWA.

This paper discusses an approach to improve plant performance and decision-making through greater utilization of plant data. The paper includes a list of best practices and provides examples where plant data was used to provide benefits to the operation of large water treatment facilities for EPCOR Water Services, Inc.

Shariff, R. A. Cudrak, Q. Zhang, and S. Stanley. 2003. Integration of Next Generation Process Control Strategies into Existing Water Treatment Plants. In *Proc. of AWWA Information Management Technology Conference, April 27-30, 2003, Santa Clara, CA.* Denver, Colo.: AWWA.

This paper describes advanced control techniques that are emerging in water treatment process control. Artificial neural networks (ANN) are a viable option for many existing WTPs because ANN modeling is based on large amounts of historical plant data which may already be available at many plants. One significant benefit of advanced control is that it allows plant-wide optimization versus unit process optimization of conventional controls. The paper presents two case studies where advanced controls were implemented.

Smith, H., R. Emanuel, M. Wehmeyer and B. Phillips. 1997. Alternative Project Delivery for Facility Automation and Information Management for a Multi-Plant, Systemwide Infrastructure Modernization. In *Proc. 1997 AWWA Computer Conference, April 13-16, 1997, Austin, Texas.* Denver, Colo.: AWWA.

Orlando Utilities Commission (OUC) Project 2000 is a $153 million program to modernize the water system. As part of the effort, the existing SCADA system will be replaced by a Facility Automation and Information Management (FAIM) system. This paper discusses how the FAIM system was implemented using the alternative delivery approaches selected for Water Project 2000.

Smith, H., R. Emanuel, B. Jacobsen, and W.L. Overbeek. 2000. Automation Strategies for Unattended Water Treatment Plant Operation; OUC's Water Project 2000 Experience. In

Proc. 1999 AWWA Information Management & Technology Conference, April 18-21, 1999, New Orleans, Louisiana. Denver, Colo.: AWWA.

The Orlando Utilities Commission (OUC) has implemented a Facility Automation and Information Management (FAIM) system to automate operation of it's 11 water treatment plants. The paper discusses OUC's experiences with determining automation criteria for unattended treatment plant operation; defining the technical implementation and reviewing successes and failures.

Smith, H., R. Emanuel, B. Jacobsen and M. Hartson. 2000. Integrating Maintenance and Operations: The Critical Information Link for Unattended Operations. In *Proc. 2000 AWWA Distribution System Symposium, September 10-13, 2000, New Orleans, Louisiana.* Denver, Colo.: AWWA.

The paper describes the Orlando Utilities Commission (OUC) Water Project 2000 which is a program to upgrade, improve and expand the water system and its technology. The paper focuses on the integration of historical, maintenance and operations information that allows a centralized system of operators to work more efficiently with field maintenance technicians.

Thomas, L., J. Kalinowski, C.Cook, and L. Verduzco. 2004. Fieldbus Instruments Support a LargeVision. *InTech,* July 2004. http://www.isa.org/InTechTemplate.cfm?Section=Article_Index1&template=/ContentManagement/ContentDisplay.cfm&ContentID=37013

This article describes the costs, benefits and challenges associated with using fieldbus technology as experienced by the Orange County Water District.

Younkin, C. and G. Huntley. 1996. Unattended Facilities Offer Competitive Advantage. In *Proc. 1996 AWWA Computer Conference, April 21-24, 1996, Chicago, Illinois.* Denver, Colo.: AWWA.

Water utilities are facing pressure to reduce costs and become more competitive. Technology advancements are enabling utilities to adapt new operational strategies including unattended operations. The paper discusses the Regional Water Authority's Lake Gaillard Water Treatment Plant which serves as an example of unattended operations. The paper also addresses workforce issues to improve productivity.

Wu, M.D. and J.C. Liu. 1999. Fuzzy Control of a Coagulation Reaction for the Treatment of High-Turbidity Water. *Aqua (Journal of Water Services Research and Technology).* 48(5):211-217.

A fuzzy logic controller (FLC) was used for the automatic control of coagulation in a laboratory scale water treatment plant. Using on-line streaming current and pH analyzers, it was demonstrated that the FLC functions satisfactorily and is robust in treating high-turbidity water.

Wastewater Industry

Bistany, A.S. 2005. Plug and Perform. *Water Environment & Technology,* 17(9):36, 7 pgs.

This article summarizes interviews with several leading experts in the field of automation and control for the wastewater industry. The article includes a discussion of automation advances in the past twenty years, describes typical levels of automation currently found in plants, identifies considerations for implementation of control systems, discusses human factors related to automation, addresses cost considerations and identifies future trends.

Black, J.M. 2004. HMI/PLC Control System Design Brought Key Reusable Technology Together to Provide an Integrated Control and Information System. In *Proc. of Water Environment Federation Annual Technical Conference & Exposition, October 2-6, 2004, New Orleans, LA.* Alexandria, VA: WEF.

The article provides examples of object based programming for HMI/PLC control systems that integrate CMMS, O&M Manual and operational information. The control system application components can be reused on other projects at a reduced cost. The integrated information system provides the information required to operate and maintain complex facilities more efficiently.

Davis, G. 1995. Wastewater Facility Upgrades Through Instrumentation. *Water Engineering & Management,* 142(10):38-40.

This article summarizes approach and benefits realized from instrumentation improvements made to a small wastewater treatment plant.

Debusscher, D., L.N. Hopkins, D. Demey, and P.A. Vanrolleghem. 2000. Determining the Potential Benefits of Controlling and Industrial Wastewater Treatment Plant. In *Proceedings 1st IWA World Water Congress. Paris, France, July 3-7, 2000.*

This paper presents a case study and methodology for evaluating the potential benefits of process control improvements at a full-scale industrial wastewater treatment plant. The methodology utilizes 1) a benchmark profile based on World Best Practice, 2) an estimation of potential benefits relative to yearly cost and 3) identifying potential performance improvement. The study does not replace the need for a more thorough analysis but can point out target areas for the analysis.

DeLaura, T.J. 2003. What is the ROI from IT Initiatives. In *Proc. of Water Environment Federation Annual Technical Conference & Exposition, October 12-15, 2003, Los Angeles, CA.* Alexandria, VA: WEF.

This article discusses the issues water and wastewater utilities should consider when determining the ROI on IT Initiatives. The paper asserts that IT initiatives should be aligned with business goals and strategies and should include master planning. Process Control Systems Data Integration is just one aspect of utilities' IT Initiatives. The article identifies a number of economic analysis tools that can be used to evaluate ROI and lists guidelines for choosing metrics and maximizing ROI.

Ekster, A. 2004. Golden Age. *Water Environment & Technology,* 16(6):62-64.

The author describes control strategies to optimize activated sludge system performance and provides case studies documenting lower operating costs.

Garrett, M.T., Jr. 1998. Instrumentation, Control and Automation Progress in the United States in the Last 24 Years. *Water Science and Technology,* 37(12):21-25.

This paper provides a historical overview and future predictions of instrumentation, control and automation as applied to wastewater treatment facilities in the United States.

Gillette, R.A. and D.S. Joslyn. 2001. *Thickening and Dewatering Processes: How to Evaluate and Implement an Automation Package.* Alexandria, VA: WERF

This report evaluates the capabilities and maintenance requirements of available automation for thickening and dewatering processes and suggests features for improvement. In addition the report evaluates solids concentration analyzers for their ability to accurately monitor different sewage solids and for their durability and calibration and maintenance requirements.

Hill, R.D., R.C. Manross, E.V. Davidson, T.M. Palmer, M.C. Ross, and S.G. Nutt. 2002. *Sensing and Control Systems: A Review of Municipal and Industrial Experiences.* Alexandria, Va: WERF

This research assessed and documented the state-of-the-art of wastewater treatment plant sensing and control systems. The project focused on the best examples of sensor application, control strategies and computerized process control in WWTPs. Survey results showed that most respondents justify installing automation systems because of cost savings, though less than 10% of the facilities surveyed had data demonstrating the savings. The research also found that facilities that do use automation do achieve great cost savings. The report includes costs and savings if they were known. The report further recommends that further research be conducted compare the costs and performance of a WWTP before and after implementation of a comprehensive sensing and control system.

Hill, R., B. Manross, and A. Manning. 2001. Assess Installed State-of-the-Art WWTP Sensing and Control Systems. In *Proc. of Water Environment Federation Annual Technical Conference & Exposition, October 13-17, 2001, Atlanta, Ga.* Alexandria, VA: WEF.

This paper summarizes a WERF-sponsored project which compiled a large body of information on successful and problematic practices in wastewater treatment control systems. Results of WWTP survey data are summarized and an example of a successful practice for a unit process is provided. Best practices include a list of instrumentation and equipment, detailed control strategy and estimated automation costs.

Hill, R. 1997. *Automated Process Control Strategies.* Water Environment Federation. Alexandria, VA.

This book presents control strategies, algorithms, and objectives for many of the common unit operations in wastewater treatment plants (WWTP). Strategies include a description of each unit operation; process and instrument diagram of each process; a list of instrumentation required for the strategies outlined, as well as descriptions of strategies that can be used for improving process performance, reducing costs and/or maintenance.

Kendricks, L.E., Jr. 1999. Reaping the Benefits of Wastewater Treatment Plant Automation. *Pollution Engineering.* 31(12):52-53.

This article describes the general benefits of automation for wastewater treatment plants and describes affordable implementation of systems for smaller public and private facilities that previously could not afford SCADA or advanced automation. General reference is made to automation costs, but the author recommends that the cost analysis include the hidden cost of not automating. Key to a successful implementation include 1) educating staff, 2) planning and budgeting, 3) selection of consulting engineer and 4) utility participation in design.

Kugelman, I. and J. Houthoofd. 2002. Optimization of Treatment Plant Operation. EPA/600/J-85/218 (NTIS PB86118486). Washington, D.C.:USEPA.

Literature review covering sixty-one citations on upgrading the operation of wastewater treatment plants. Topics include management, operation, maintenance, and training; process control and modeling; instrumentation and automation; and energy savings.

LaMontagne, P.L. 2004. Degrees of Automation. 2004. *Water Environment & Technology,*16(11):43-45.

This article discusses how the operation control needs are quite different for large plants and small plants. Citing single examples in both cases, the author generalizes that full automation is best where there is no central control room or where the number of control points would burden the central operator.

Lant, P. and M. Steffens. 1998. Benchmarking for Process Control: "Should I Invest in Improved Process Control?". *Water Science and Technology.* 37(12):49-54.

A benchmarking procedure is used to gain support for investment in process control projects at wastewater treatment facilities. Results of the benchmarking can identify good candidates for a thorough process control cost/benefit analysis. Results of surveys found that most benefit for WWTP can be gained through savings in deferred capital expenditures and operating costs. It was noted that approximately 60% of the plants were capacity limited but the cost of being at capacity or the benefit of increasing throughput are generally not known.

O'Brien, A., Swanback, S., and Marks, K. 2000. Visioning a New California Wastewater Treatment Plant Unattended Operation/Unrestricted Effluent Use. 2000. In *Proc. of Water Environment Federation Annual Technical Conference & Exposition, October 14-18, 2000, Anaheim, CA.* Alexandria, VA: WEF.

This paper discusses several aspects of developing and realizing a vision for the construction of a new WWTP in California. Of most interest is the consideration of unattended operation. The utility ultimately defined "unattended" as a small operations staff assigned for the week day shift only with no staff on the off shifts or weekends. Planning for "unattended" operations led to a realization that the best selection for process and equipment was often based on operational simplicity or equipment reliability. In addition all pumping and chemical feed systems, including backup systems, were automated.

Olsson, G., M. Nielsen, Z. Yuan, A. Lynggaard-Jensen, J-P. Steyer. 2005. *Instrumentation, Control and Automation in Wastewater Systems.* London: IWA Publishing.

This book summarizes the state-of-the-art of instrumentation, control and automation and its applications in wastewater treatment systems. The book focuses on how technology can be used for better operation. Economic benefits of different control and operations alternatives are quantified. The book includes several case studies showing how automation have improved costs, operation and robustness of WWTP operation.

Olsson, G. and P. Ingildsen. 2003. Automation in Wastewater Treatment Plants. In *Proc. of Water Environment Federation Annual Technical Conference & Exposition, October 12-15, 2003, Los Angeles, CA.* Alexandria, VA: WEF.

This article provides an overview of the state of automation in the wastewater treatment industry. It provides a discussion on the level of automation typically utilized, automation trends, benefits of automation, barriers or constraints and examples of specific process control strategies typically used.

Patrick, R., J. Rompala, A. Symkowski, W. Kingdom, R. Serpente, N. Freeman, B. Stevens, C. Koch, and T. Kochaba. 1997. *Benchmarking Wastewater Operations – Collection, Treatment, and Biosolids Management.* Alexandria, Va.: WERF.

Developed cost models and benchmarking for wastewater systems utilizing surveys and case studies to gather information. Direct cost benefits of automation are not quantified, however, the research found a statistically significant correlation between increased process automation and lower cost operation. In addition, a case study for the City of Anchorage Water and Wastewater System showed that over a 10 year period, plant size grew by 62% while staff was reduced by 15% (47 people). The reduction was attributed to three factors, one of which was automation.

Pramanik, A., P. LaMontagne, and P. Brady. 2002. Automatic Improvements. *Water Environment & Technology*. 14(10):46-50.

The article provides an overview of key considerations in automating sludge thickening and dewatering processes to optimize performance and lower operating costs. The article summarizes results of the 2001 WERF report "Thickening and Dewatering Processes: How to Evaluate and Implement an Automation Package".

Quick, G., R. Emanuel, J. O'Connor. 1997. City of Austin's Control System Modernization Yields Operational Benefits for the South Austin Regional Wastewater Treatment Plant. In *Proc. of Water Environment Federation Annual Technical Conference & Exposition, October 18-22, 1997, Chicago, IL.* Alexandria, VA: WEF.

This paper highlights the system architecture and project delivery method for instrumentation and control improvements at the City of Austin's South Austin Regional WWTP. Benefits from the improvements are identified and include specific energy cost savings data realized from improved blower control. Risk mitigation included dual redundant DCUs, dual redundant fiber optic data network and dual mirrored workstations. Other risk mitigation approaches included 1)detailed installation design in lieu of functional descriptions to minimize change orders, 2) evaluated proposals for equipment and suppliers to ensure quality and 3) application software development by the design engineer to ensure continuity of project objectives.

Ross, B., B. Brunner, L. Mincy, and G. Jones. 2003. Reinventing SCADA in the 21st Century. In *Proc. of Water Environment Federation Annual Technical Conference & Exposition, October 12-15, 2003, Los Angeles, CA.* Alexandria, VA: WEF.

This article provides a detailed description of the SCADA system improvements made for the City of Lansing, MI wastewater facilities. The article includes a listing of new instrumentation and equipment installed for process control and monitoring, MAN architecture development, integration of security cameras for process control and safety, and development of web-based O&M manuals. Benefits of the SCADA system improvements are also discussed.

Russo, P. 2001. District Wide Conversion to Unmanned Weekends and Nights. In *Proc. of Water Environment Federation Annual Technical Conference & Exposition, October 13-17, 2001, Atlanta, Ga.* Alexandria, VA: WEF.

This paper summarizes one utilities approach to converting to partially unattended operations for three wastewater treatment facilities for the North Shore Sanitary District. The paper describes the approach and methodology, the required levels of automation, staffing considerations, savings and costs. The plant automation upgrades needed to support unattended weekend and night shifts had a five year pay back period.

Wensloff, David A. 1998. Optimizing Your Industrial Wastestream Costs. 1998. *Water/Engineering & Management*, 145(3):26.

Paper discusses how Wastewater utilities can look at surcharges for industrial sources to cover the costs of treating the associated wastes. The paper includes a discussion on chemical usage evaluation and optimization, including optimizing coagulation chemical costs. Paper provides a general discussion on the topics, and does not provide specific costs or benefit information, or discuss unattended operation.

Electric Utility Industry

Al-Sum, E.A., A. Sattar, and M. Abdul Aziz. 1993. Automation of Water Treatment Plants and its Application in Power and Desalination Plants. *Desalination.* 92(1993): 309-321.

This paper discusses specific applications for chemical treatment automation in power and desalination plants. Measuring principles for primary sensors, control loop descriptions and advantages and disadvantages of each are discussed.

Brown, D.L., J.W. Skeen, P. Daryani, and F.A. Rahimi. 1991. Prospects for Distribution Automation at Pacific Gas & Electric Company. *IEEE Transactions on Power Delivery*, 6(4):1946-1954, October 1991.

This paper presented a method to evaluate the feasibility of distribution automation utilizing a computer based model and standard algorithms based on present value economic analysis. The method was applied to two case studies for PG&E facilities. The results of the study showed that substation automation could be justified solely on lower operation and maintenance costs.

Cassel, W.R. 1993. Distribution Management Systems: Functions and Payback. *IEEE Transactions on Power Systems,* 8(3):796-801.

This paper describes the application of a Distribution Management System which includes SCADA, distribution automation, feeder automation, GIS, customer information and energy management systems. The paper discusses payback opportunities and cost/benefit analysis methodology.

Dasch, J. 2001. Retrofitted State-of-the-Art Control Systems Improve Plant Performance. *Power Engineering,* March 2001.

This article describes partnering relationship between power utilities and process control system suppliers to implement control system improvements and replacements. Three example projects are described.

Dondi, P., Y. Peeters, and N. Singh. 2001. Achieving Real Benefits by Distribution Automation Solutions. In *Proc. of CIRED2001 International Conference and Exhibition on Electricity Distribution, June 18-21, 2001, Amsterdam, The Netherlands.* IEE.

The paper describes the results of using new tools to extend the traditional planning tools with financial impacts of implementing network changes and automation strategies. Case studies demonstrate where automation solutions bring the best cost-benefit for the utility.

Gruenemeyer, D. 1991. Distribution Automation: How Should it be Evaluated? In *Proc. of Rural Electric Power Conference, April 28-30, 1991, Dearborn, MI.* IEEE.

This paper describes the benefits that can be realized from distribution automation systems, identifies typical costs of automation and describes a cost/benefit analysis methodology for automation projects.

Haacke, S., S. Border, D. Stevens and B. Uluski. 2003. Plan Ahead for Substation Automation. *IEEE Power & Energy Magazine,* 1(2):32-41.

This paper describes a methodology for developing a business case for substation automation projects.

Lehtonen, M. and S. Kupari. 1995. A Method for Cost Benefit analysis of Distribution Automation. In *Proc. of EMPD '95 International Conference on Energy Management and Power Delivery, November 21-23, 1995,* 1:49-54. IEEE.

This paper provides a computer-based approach for cost benefit analysis of distribution automation. This is an example of the electric utilities' advancements in developing detailed and systematic approaches to economic analysis for automation projects.

Morris, J.F., F.J. Kern, and E.F. Richards. 1988. Distribution Automation of the Association of Missouri Electric Cooperatives – A Statewide Evaluation of Load Management. *IEEE Transactions on Industry Applications,* 24(5):782-791, September/October 1988.

This paper presents a technical and economic feasibility study for a state-wide distribution automation system. The study includes quantified benefits and costs and a cost/benefit analysis for distribution automation technology alternatives.

Newton, C. 1995. Benefits Analysis for Automation Programs. *Transmission & Distribution.* 47(7):12.

This article discusses the need to include non-financial costs and benefits in justification of automation projects for transmission and distribution systems.

Rahimi, A.F., L.P. Hajdu, L. Kiss, and L. Balogh. 1993. A General Cost-Benefit Analysis Methodology for the Evaluation of EMS/SCADA Procurement Alternatives. In *Proc. IEEE/NTUA Athens Power Tech Conference, September 5-8, 1993, Athens, Greece.*

This paper provides a cost/benefit economic analysis methodology based on comparison of each alternative with a "status quo". Quantifiable benefits and costs are identified as well as unquantifiable benefits. An example application is presented.

Roberts, G.V. 1999. Analysis of Reliability of Networks and Justification of Automation. In *IEE Colloquium on Remote Control and Automation on 11kV Networks Beyond the Primary Substation (Ref. No. 1999/195), November 22, 1999*. IEE.

This paper asserts that financial justification for automation schemes or other customer performance improvement measures are difficult to produce. Recent improvements in software solutions are now available to carry out the cost benefit analysis of electricity network automation schemes and other methods of improving performance. The results are that investment options can be analyzed in a quantitative and auditable manner, however automation may be low in the ranking.

Hydroelectric Industry

Benson, B. 2005. Beyond Automation. *International Water Power & Dam Construction,* 57(6):30-33.

This article summarizes a feasibility study for the US Army Corps of Engineers which investigated the costs and benefits to evaluate options for automation, staffing levels and responsibilities at six hydroelectric plants along the Missouri River. Levels of automation included unattended operations with the economic analysis was based on payback. The article also summarized risk issues for unattended plants.

Clemen, D.M., G. Llort, D. Augustine, and W. Hindsley. 1997. Control Automation of NIPSCO Hydroelectric Plants. 1997. In *Proc. of Waterpower '97, August 5-8, 1997, Atlanta, GA.* ASCE.

This article summarizes the technical requirements and implementation approach for a remote-unattended control system for two hydroelectric plants owned by Northern Indiana Public Service Company (NIPSCO). System requirements included redundant CPUs in the RTU, dual databases, multiple communication ports, low cost and maintainability. The implementation approach included selection of a vendor based on the system requirements, development of detailed specifications, detailed drawing submittal review, FAT/SAT testing and availability demonstration.

Duvall, M. 1999. Upgraded SCADA System Gives Hydro Plant Greater Reliability and Room to Grow. *Power Engineering*, October 1999, pgs 49-51.

This paper describes SCADA system architectural changes and benefits gained by upgrading Virginia Power's Bath County Power Station hydroelectric plant SCADA system. The upgrade project re-used existing I/O systems, but replaced servers, application software, operating system and communications network. The project extended the longevity of the plant's data system 10 to 15 years and provided increased flexibility, functionality, scalability and expansion capabilities. The author also speculates on where SCADA systems are headed in the future including standardized protocols and integration with geographic information systems (GIS).

EPRI (Electric Power Research Institute). 1989. *Hydropower Plant Modernization Guide, Volume 3: Automation.* EPRI GS-6419. Palo Alto, Ca: EPRI.

This guide provides procedures for hydroelectric utilities to identify plants that are potentially suitable for cost-effective implementation of automation. The approach includes initial screening, feasibility review of systems for a typical plant, and a listing of instrumentation and other equipment required for each system including ranges of cost. Requirements for different levels of automation are addressed including semi- and fully-automatic, remotely controlled and unmanned sites. The guide includes a detailed economic evaluation using a present worth analysis over a 10 year period.

IEEE (Institute of Electrical and Electronics Engineers). 1996. *Guide for Computer-Based Control for Hydroelectric Power Plant Automation.* IEEE Std 1249-1996. New York: IEEE. Approved December 10, 1996.

This guide addresses application, design concepts and implementation of computer-based control systems for hydroelectric power plant automation. Functional capabilities, performance and interface requirements, hardware considerations, system testing and acceptance are discussed. Case studies are also presented.

Terry, W.W. 2002. Hydro Automation Program Improves Efficiency and Reduces Operating Expense at TVA. *Power Engineering.* 106(3):54-60.

Tennessee Valley Authority (TVA) plans to completely automate the operation of all 29 of its conventional hydropower plants over an eight year period. TVA performed a study to evaluate the costs of implementing complete automation compared to the economic benefits. The economic analysis showed a 30% internal rate of return based on a cost of $50 million and a cost savings for $58.9 million over the eight year period. The paper further describes the general control philosophy, expected benefits and phased implementation approach.

Irrigation

Latimer, E.A., and D.L. Reddell. 1990. Components for an Advanced Rate Feedback Irrigation System (ARFIS). *Transactions of the American Society of Agricultural Engineers.* 33(4):1162-1170.

This article presented experimental control strategies and components needed for advanced automation of an irrigation system. A cost analysis and economic evaluation was completed which indicated that irrigation automation could provide a favorable net return. The article included recommendations for future studies and discussion of potential improvements in communication technologies that could lower capital costs.

Maskey, R., G. Roberts, and B. Graetz. 2001. Farmers' Attitudes To The Benefits And Barriers Of Adopting Automation For Surface Irrigation On Dairy Farms In Australia. *Irrigation and Drainage Systems.* 15:39-51.

This article discusses barriers and benefits to automation as it relates to irrigation for dairy farms. The results indicate that the most important influence on the level of automation is cost.

U.S. Government References

U.S. Department of Energy, Office of Energy Efficiency and Renewable Energy, Federal Energy Management Program. 1996. 10 CFR 436, Subpart A, *Methodology and Procedures for Life Cycle Cost Analyses.* http://www.access.gpo.gov/nara/cfr/waisidx_04/10cfr436_04.html

This document sets forth rules to promote life cycle cost effective investments in building energy and water systems and energy and water conservation projects.

U.S. Department of Commerce, Office of the Assistant Secretary for Conservation and Renewable Energy, Federal Energy Management Program, NIST Handbook 135, 1995 Edition. *Life Cycle Costing Manual for the Federal Energy Management Program.* http://www.bfrl.nist.gov/oae/publications/handbooks/135.pdf

This handbook is a guide to understanding Federal Energy Management Program (FEMP) life-cycle cost (LCC) methodology for evaluation of building energy and water systems and energy and water conservation projects as outlined in FEMP rules published in 10 CFR 436, Subpart A. A supplement to the handbook, *Energy Price Indices and Discount Factors for LCC Analysis*, NISTIR 85-3273-X is published annually to provide current discount rates and factors required for LCC analysis.

U.S. Environmental Protection Agency. 2000. *Community Water System Survey 2000.* EPA-815-R-02-005A. http://www.epa.gov/safewater/cwssvr.html

The U.S. Environmental Protection Agency (EPA) conducted the 2000 Community Water System (CWS) Survey to obtain data to support its development and evaluation of drinking water regulations. The 2000 survey, which represents the fifth edition since 1976, incorporated responses from 1,246 water systems. This survey includes information relating to types of plants and processes, percentages of systems that run attended 24/7, as well as percentages of systems that utilize SCADA Systems, among many other categories.

U.S. Environmental Protection Agency. 2001. *1999 Drinking Water Needs Survey, Modeling the Cost of Infrastructure.* EPA 816-R-01-005.
http://www.epa.gov/ogwdw/needs/fullcost.pdf

Provides the results of a comprehensive data gathering effort done by the USEPA on the costs associated with water system infrastructure. One section of the report presented the results from the cost data gathered for new and rehabilitation, water treatment facility SCADA and automation projects.

U.S. Environmental Protection Agency. 2003. *Small Drinking Water Systems Handbook: A Guide to "Packaged" Filtration and Disinfection Technologies with Remote Monitoring and Control Tools*, May 2003. EPA-600-R-03-041.
http://www.epa.gov/ord/NRMRL/Pubs/600R03041/600R03041.pdf

This document provides a detailed overview of water treatment for small systems. It includes discussion on contaminants, common water supply problems and solutions, regulatory overview, treatment technologies, remote control and monitoring discussions, funding and technical resources, among other topics. Of relevance to this AWWARF project, it provides an outline of the applicability of automation and remote monitoring to different treatment processes.

U.S. Environmental Protection Agency. 2005. *Drinking Water Infrastructure Needs Survey and Assessment, Third Report to Congress.* EPA 816-R-05-001. http://www.epa.gov/ogwdw000/needssurvey/pdfs/2003/report_needssurvey_2003.pdf

Report identifies the total national need for water system infrastructure as $276.8 billion over the next 20 years which includes an estimated $2.3 billion for computer and automation equipment, system security and emergency power generators.

U.S. Federal Government, Office of Management and Budget. 1992. Memorandum for Heads of Executive Departments and Establishments. *Guidelines and Discount Rates for Benefit-Cost Analysis of Federal Programs.* Circular No. A-94 Revised (Transmittal Memo No. 64). http://www.whitehouse.gov/omb/circulars/a094/a094.html.

Provides the Federal guidelines for developing cost-benefit analysis for federal projects. It shows the preferred method includes net present value analysis.

U.S. General Accounting Office. 2004. *Water Infrastructure Comprehensive Asset Management Has Potential to Help Utilities Better Identify Needs and Plan Future Investments.* GAO-04-461. Washington, D.C.: GAO. http://www.gao.gov/new.items/d04461.pdf

Comprehensive report by the U.S. General Accounting Office which included interviews with U.S., Australian and New Zealand utilities that have implemented asset management plans. The study also included interviews with water industry associations and the EPA. The report concludes that asset management allows utilities to realize benefits from 1) improved decision making because they have more accurate information about their capital assets and 2) more productive relationships with ratepayers, governing authorities and other stakeholders. The study also concluded that the EPA can play a stronger role in encouraging water utilities to use asset management. The EPA should establish a central repository for information on asset management. Challenges to mandating asset managements include defining an adequate asset management plan and the ability of states to oversee and enforce compliance.

State Government References

California Department of Health. California Safe Drinking Water Act & Related Laws and Regulations. Title 22, Division 4, Chapter 15, Articles 2,3,4,5 and 16. http://www.dhs.ca.gov/ps/ddwem/publications/lawbook/dwregulations-02-27-06.pdf

Florida Administrative Code. Department of Environmental Protection. Chapters 62-550, Drinking Water Standards Monitoring and Reporting, and Chapter 62-699, Treatment Plant Classification and Staffing. http://fac.dos.state.fl.us/faconline/chapter62.pdf

Illinois Administrative Code. Title 35, Subtitle F, Public Water Supplies, Chapter I, Pollution Control Board, Part 603, Ownership and Responsible Personnel, and Part 611, Primary Drinking Water Standards, Subparts L, M, N, O, Q. http://www.ipcb.state.il.us/SLR/IPCBandIEPAEnvironmentalRegulations-Title35.asp

Michigan Department of Environmental Quality, Water Division, Supplying Water to the Public, Part 7; R325.10701-R325.10738, Surveillance, Inspection and Monitoring and Part 9 R325.11901-R325.11918, Examination and Certification of Operators. http://www.michigan.gov/deq/0,1607,7-135-3307_4132-14902--,00.html

New York State Department of Health, Title 10, Part 5, Subpart 5-1, Public Water Systems. http://www.health.state.ny.us/nysdoh/water/part5/subpart5.htm

Ohio Administrative Code, Part 3745-81, Primary Drinking Water Rules and Part 3745-7-04 (Draft), Treatment works and sewerage system classification and staffing requirements. http://www.epa.state.oh.us/ddagw/oac.html#374507

Pennsylvania Code, Title 25, Environmental Protection, Chapter 109, Safe Drinking Water, Subchapter C, Monitoring Requirements and Subchapter G, System Management Responsibilities. http://www.pacode.com/secure/data/025/025toc.html

Texas Administrative Code, Title 30, Environmental Quality, Part 1, Chapter 290, Public Drinking Water, Subchapter D, Rules and Regulations for Public Water Systems, and Subchapter

F, Drinking Water Standards Governing Drinking Water Quality and Reporting Requirements for Public Water Systems.
http://info.sos.state.tx.us/pls/pub/readtac$ext.ViewTAC?tac_view=4&ti=30&pt=1&ch=290

Business Information Technology

Kaplan, R.S., and D.P. Norton. 1996. *The Balanced Scorecard.* Boston, MA: Harvard Business School Press.

Provides guidelines and methodologies for evaluating projects that include tangible and significant intangible costs and benefits.

Petrochemical Industry

Bonavita, N. 2001. Benchmarking of Advanced Technologies for Process Control: An Industrial Perspective. *In Proceedings of the American Control Conference, June 25-27, 2001, Arlington, VA.* Dayton, OH: AACC.

This paper summarizes process automation benchmarking for industrial plants such as refineries and petrochemical plants. Improved product quality and reduced costs are major drivers for industries to embrace advanced automation technologies, however the technologies must satisfy the expected ROI required by management.

Bozenhardt, H. 1987. Analyzing the Benefits of Automation: A Methodology. Paper #87-1008 1-55617-054: 73-83. Research Triangle Park, NC: ISA.

This paper discusses a methodology to justify automation projects using a chemical plant as an example. The methodology is based on a present value economic analysis which includes quantifiable, total life cycle costs and credits including capital expense, deferred capital expense, project expense (i.e. upgrade HVAC), credits (i.e. energy savings), tax credits, depreciation, maintenance (savings or costs), design, implementation, operating profit and taxes. The resulting cash flow chart is analyzed based on DCFR and payback.

Collins, K., S. Treiber, and J. Walker. 1997. Multivariable Control Benefits Alberta Plant. *Oil & Gas Journal,* 95(18): 118-122.

This article describes the multivariable control strategies used to reduce product variability at a gas plant resulting in a project payback period of less than one year.

General/Other

Brisk, M.L. 2004. Process Control: Potential Benefits and Wasted Opportunities. In *Proc. of 5th Asian Control Conference, July 2004, Melbourne, Australia.*

This paper first summarizes the 1988 University of Sydney Warren Centre study which developed a systematic approach to determine the potential benefits of improved automation and also discusses the potential benefits and sites case study examples of cost savings realized. It further investigates the missed opportunities to gain benefits from optimized process control.

Carr, P.G., P.S. Beyor. 2005. Design Fees, the State of the Profession, and a Time for Corrective Action. *Jour. of Management in Engineering.* 21(3):110-117.

This article discusses the history and practice of using percentage of construction as a basis for determining professional design fees. Because professional service fees have not been adjusted uniformly for inflation it has resulted in a significant decay in fees, when inflation is taken into account. The article proposes using a family of fee schedules based on project complexity that are adjusted over time to reflect inflationary impacts to construction and professional services.

Dorresteijn, R.C., G. Wieten, P.T.E. van Santen, M.C. Philippi, C.D. de Gooijer, J. Tramper and E.C. Beuvery. 1997. Current Good Manufacturing Practice in Plant Automation of Biological Production Process. *Cytotechnology.* 23:19-28.

Current good manufacturing practices (cGMP) are developed by the U.S. Food and Drug Administration for the production of biologicals. This articles discusses current trends in the production of biologicals toward complete automation to meet cGMP as well as process optimization to increase consistency and reproducibility. This article provides an example of how another industry uses automation to meet strict government regulations.

Marlin, T.E., M.L. Brisk, G.W. Barton, and J.D. Perkins. 2001. Experiences from an Industry-University Study on Control Benefits. *Pulp & Paper Canada*, 102(4):34-37.

This paper summarizes the 1988 University of Sydney Warren Centre study which developed a systematic approach to determine the potential benefits of improved automation in process industries and presents developments since the study was completed. The general method concentrates on plant deviations from performance goals, identifies methods for reducing the deviations, develops a "long list" of potential items and refines the opportunities to a "short list". The total benefits are calculated based on the individual contributions for each opportunity.

Singh, A. 2001. An Economic Feasibility Analysis for the Automation of Hazardous Waste Landfills. *Cost Engineering,* 35(4):13-21.

This paper outlines the automation improvements for remote operation of hazardous waste landfills and presents a detailed economic analysis based on present worth cost and incremental rate of return (IROR) which justifies implementation of the automation improvements.

REFERENCES

Al-Sum, E.A., A. Sattar, and M. Abdul Aziz. 1993. Automation of Water Treatment Plants and its Application in Power and Desalination Plants. *Desalination.* 92(1993): 309-321.

Arora, H; LeChevallier, M; Barrer, P. 1996. Energy Management Opportunities in the American Water System. In *Proc. AWWA Annual Conference and Exposition, June 26, 1996, Toronto, Canada.* Denver, Colo: AWWA.

AWWA (American Water Works Association). 2000. *Operational Control of Coagulation and Filtration Processes*, AWWA Manual M37, Second Edition. Denver, Colo.: AWWA.

AWWA (American Water Works Association). 2001. *Instrumentation and Control,* AWWA Manual M2, Third Edition. Denver, Colo.: AWWA.

AWWA (American Water Works Association). 2005. *Water Treatment Plant Design, Fourth Edition.* New York: McGraw-Hill.

AwwaRF (American Water Works Association Research Foundation). 2005. Case Study: Energy and Water Quality Management System (EWQMS) Saves Electricity Dollars. http://www.awwarf.org/research/TopicsAndProjects/Resources/caseStudies/caseStudyEWQMS.aspx

Barnes, M., C. Brophy and R.J. Daly. 1997. Automating the Lake DeForest Treatment Plant: A Step-Wise Approach to Unattended Operations. In *Proc. 1997 AWWA Computer Conference, April 13-16, 1997, Austin, Texas.* Denver, Colo.: AWWA.

Baumeister, R.A. 1996. Treatment Plant Automation: Implications for Treatment Optimization. In *Proc. 1996 AWWA Annual Conference and Exposition, June 23-27, 1996, Toronto, Ontario, Canada.* Denver, Colo: AWWA.

Baxter, C.W., Zhang, Q., Stanley, S.J, Shariff, R., Tupas, R-R.T, 2001. Drinking Water Quality and Treatment: the use of artificial neural networks. *Canadian Journal of Civil Engineering,* 28: 26-35.

Benson, B. 2005. Beyond Automation. *International Water Power & Dam Construction,* 57(6):30-33.

Bevan, D., C. Cox, and A. Adgar. 1998. Implementation Issues when Installing Control and Condition Monitoring at Water Treatment Works. *IEE Colloquium on Industrial Automation and Control: Distributed Control for Automation (Digest No. 1998/297), March 4, 1998, London, UK.* IEE.

Bistany, A.S. 2005. Plug and Perform. *Water Environment & Technology*, 17(9):36-43.

Black, J.M. 2004. HMI/PLC Control System Design Brought Key Reusable Technology Together to Provide an Integrated Control and Information System. In *Proc. of Water Environment Federation Annual Technical Conference & Exposition, October 2-6, 2004, New Orleans, LA.* Alexandria, VA: WEF.

Boman, E. 2003. Case Study 2: Town of Fairfield Water Pollution Control Facility Upgrade. Presented at New Haven Energy Workshop for Water & Wastewater Facilities -- Clean Water, Clear Skies, and Cost Savings: What Drinking Water and Wastewater Utilities Can Learn From Energy Efficiency Pioneers in Connecticut, November 20, 2003. http://www.summersaver.state.ct.us/documents.htm

Bonavita, N. 2001. Benchmarking of Advanced Technologies for Process Control: An Industrial Perspective. In *Proc. Of the American Control Conference, June 25-27, 2001, Arlington, VA.* Dayton, OH: AACC.

Bozenhardt, H. 1987. *Analyzing the Benefits of Automation: A Methodology.* Paper #87-1008 1-55617-054: 73-83. Research Triangle Park, NC: ISA.

Brisk, M.L. 2004. Process Control: Potential Benefits and Wasted Opportunities. In *Proc. Of 5th Asian Control Conference, July 2004, Melbourne, Australia.*

Brooks, R.L. 1992. Operational Cost Savings from SCADA Systems – Fact or Fiction? In *Proc. Of 1992 AWWA Computer Conference, April 12-15, 1992, Nashville, Tennessee.* Denver, Colo.: AWWA.

Brown, D.L., J.W. Skeen, P. Daryani, and F.A. Rahimi. 1991. Prospects for Distribution Automation at Pacific Gas & Electric Company. *IEEE Transactions on Power Delivery*, 6(4):1946-1954, October 1991.

Brown, J. 1989. Control System Conserves Staff Time, Saves Maintenance Costs. *Waterworld News,* 5(3):32.

California Department of Health. California Safe Drinking Water Act & Related Laws and Regulations. Title 22, Division 4, Chapter 15, Articles 2,3,4,5 and 16. http://www.dhs.ca.gov/ps/ddwem/publications/lawbook/dwregulations-02-27-06.pdf

California Energy Commission. 2003a. Electric Load Management. [online] Available: http://www.energy.ca.gov/process/pubs/eload.pdf

California Energy Commission. 2003c. Success Story: San Juan Water District N. Peterson Water Treatment Plant. [online] Available: http://www.energy.ca.gov/process/pubs/sanjuan.pdf

California Energy Commission. Success Story: Madera Valley Water Company. [online] Available: http://www.energy.ca.gov/process/pubs/madera.pdf

California Energy Commission. 2003b. Success Story: Moulton Niguel Water District. [online] Available: http://www.energy.ca.gov/process/pubs/moulton.pdf

Carns, Keith. 2004. Bringing Energy Efficiency to the Water and Wastewater Industry: How Do We Get There? Presented at the ACEEE Water & Wastewater Energy Roadmap Workshop, Washington D.C., July 29, 2004. http://www.aceee.org/industry/carns.pdf

Carr, P.G., P.S. Beyor. 2005. Design Fees, the State of the Profession, and a Time for Corrective Action. *Jour. Of Management in Engineering.* 21(3):110-117.

Cascos, G. and T. DeLaura. 2003. Improving Energy Efficiency Using a Department-Wide SCADA System. In *Proc. AWWA Annual Conference and Exposition, June 15-19, 2003, Anaheim, CA.* Denver, Colo: AWWA.

Cassel, W.R. 1993. Distribution Management Systems: Functions and Payback. *IEEE Transactions on Power Systems*, 8(3):796-801.

Chaudhry, S. 2005. Utility Strategies For Shifting Peak Demand Period Water & Energy Use Regional Strategies: Peak Demand Gap & Critical Peak Pricing. Presented at California Energy Commission Energy Workshops for Water and Wastewater Agencies. August 2005.http://www.energy.ca.gov/process/water/water_pubs/2005-08-25_CHAUDHRY_DWR_WORKSHOP.PDF

Clemen, D.M., G. Llort, D. Augustine, and W. Hindsley. 1997. Control Automation of NIPSCO Hydroelectric Plants. 1997. In *Proc. Of Waterpower '97, August 5-8, 1997, Atlanta, GA.* Reston, V.A.: ASCE.

Collins, K., S. Treiber, and J. Walker. 1997. Multivariable Control Benefits Alberta Plant. *Oil & Gas Journal*, 95(18): 118-122.

Copithorn, R.R. and J.K. Warrick. 1989. Automation of a Water Treatment Plant for Unattended Operation and Shutdown. In *Proc. AWWA Computer Specialty Conference, April 2-4, 1989, Denver, Colorado.* Denver, Colo: AWWA.

Covelli, D. 2003. Nation's Largest AMR System Comes Online in DC. *Water & Wastes Digest,* 43(7).

Critchley, R.F., E.O. Smith and P. Pettit. 1990. Automatic Coagulation Control at Water Treatment Plants in the North-West Region of England. *Journal of the Institution of Water and Environmental Management,* 4(6):535-543.

Dasch, J. 2001. Retrofitted State-of-the-Art Control Systems Improve Plant Performance. *Power Engineering*, March 2001.

Davis, G. 1995. Wastewater Facility Upgrades Through Instrumentation. *Water Engineering & Management*, 142(10):38-40.

Debusscher, D., L.N. Hopkins, D. Demey, and P.A. Vanrolleghem. 2000. Determining the Potential Benefits of Controlling and Industrial Wastewater Treatment Plant. In *Proceedings 1st IWA World Water Congress. Paris, France, July 3-7, 2000.*

DeLaura, T.J. 2003. What is the ROI from IT Initiatives. In *Proc. of Water Environment Federation Annual Technical Conference & Exposition, October 12-15, 2003, Los Angeles, CA.* Alexandria, VA: WEF.

Dentel, S.K., and K.M. Kingery. 1988. *An Evaluation of Streaming Current Detectors*. Denver, Colo.: AwwaRF and AWWA.

Dondi, P., Y. Peeters, and N. Singh. 2001. Achieving Real Benefits by Distribution Automation Solutions. In *Proc. of CIRED2001 International Conference and Exhibition on Electricity Distribution, June 18-21, 2001, Amsterdam, The Netherlands*. IEE.

Dorresteijn, R.C., G. Wieten, P.T.E. van Santen, M.C. Philippi, C.D. de Gooijer, J. Tramper and E.C. Beuvery. 1997. Current Good Manufacturing Practice in Plant Automation of Biological Production Process. *Cytotechnology*. 23:19-28.

Duvall, M. 1999. Upgraded SCADA System Gives Hydro Plant Greater Reliability and Room to Grow. *Power Engineering*, October 1999, pgs 49-51.

Ebeling, Charles E., 1997, *An Introduction to Reliability and Maintainability Engineering*, McGraw-Hill Companies, Inc., Boston.

Ehlen, Dana J., Mark Maxwell, and John Schmitz. 1992. Custom-Blended Coagulant Reduces Costs, Improves Water Quality. *Water Engineering & Management*, 139(12):17.

Ekster, A. 2004. Golden Age. *Water Environment & Technology*, 16(6):62-64.

Emanuel, R.C., and B. Beaudet. 2003. Designing Automation Facilities. In *Proc. AWWA Annual Conference and Exposition, June 15-19, 2003, Anaheim, CA.* Denver, Colo: AWWA.

EPRI (Electric Power Research Institute). 1989. *Hydropower Plant Modernization Guide, Volume 3: Automation*. EPRI GS-6419. Palo Alto, Ca: EPRI.

EPRI (Electric Power Research Institute). 1994. *Energy Audit Manual for Water/Wastewater Facilities: A Guide for Electric Utilities to Understanding Specific Unit Processes and Their Energy/Demand Relationships at Water and Wastewater Plants*. CR-104300. Palo Alto, Calif.: EPRI.

EPRI (Electric Power Research Institute). 1996. *Water and Wastewater Industries: Characteristics and Energy Management Opportunities*. CR-106941. Palo Alto, Calif.: EPRI.

EPRI (Electric Power Research Institute). 1997. *Quality Energy Efficiency Retrofits for Water Systems.* CR-107838. Palo Alto, Calif.: EPRI.

EPRI (Electric Power Research Institute). 2001. *Summary Report for California Energy Commission Energy Efficiency Studies.* Palo Alto, Calif.: EPRI.

Florida Administrative Code. Department of Environmental Protection. Chapters 62-550, Drinking Water Standards Monitoring and Reporting, and Chapter 62-699, Treatment Plant Classification and Staffing. http://fac.dos.state.fl.us/faconline/chapter62.pdf

Freeman, I.C., and S.J. Prutz. 2004. Best Practices for the Reduction of SCADA System Total Lifecycle Cost. In *Proc. AWWA Annual Conference and Exposition, June 13-17, 2004, Orlando, FL.* Denver, Colo: AWWA.

Frey, M. and L. Sullivan. 2004. *Practical Application of Online Monitoring.* Denver, Colo.:AwwaRF and AWWA.

Garrett, M.T., Jr. 1998. Instrumentation, Control and Automation Progress in the United States in the Last 24 Years. *Water Science and Technology*, 37(12):21-25.

Gillette, R.A. and D.S. Joslyn. 2001. *Thickening and Dewatering Processes: How to Evaluate and Implement an Automation Package.* Alexandria, Va: WERF

Goethert, W., M. Fisher. 2003. Deriving Enterprise-Based Measures Using the Balanced Scorecard and Goal Driven Measurement Techniques. Software Engineering Institute. Carnegie Mellon University.
http://www.sei.cmu.edu/pub/documents/03.reports/pdf/03tn024.pdf

Gotoh, K., J.K. Jacobs, S. Hosoda, and R.L. Gertsberger, eds. 1993. *Instrumentation and Computer Integration of Water Utility Operations.* Denver, Colo.: AwwaRF.

Graney, G.D. 2000. Optimizing the "True Cost of Ownership" for SCADA and Automation Solutions Supporting Water Treatment Plants. In *Proc. 2000 AWWA Information Management & Technology Conference, April 16-19, 2000, Seattle, Washington.* Denver, Colo.: AWWA.

Great Lakes – Upper Mississippi River Board of State and Provincial Public Health and Environmental Managers (Ten State Standards). 2003. *Recommended Standards for Water Works, Policy Statement on Automated/Unattended Operation of Surface water Treatment Plants.* Albany, NY: Great Lakes – Upper Mississippi River Board of State and Provincial Public Health and Environmental Managers.

Groshong, N., General Manager. Umpqua Basin Water Association, Inc., Personal Interview. January 18, 2006.

Gruenemeyer, D. 1991. Distribution Automation: How Should it be Evaluated? In *Proc. of Rural Electric Power Conference, April 28-30, 1991, Dearborn, MI.* New York, N.Y.: IEEE.

Haacke, S., S. Border, D. Stevens and B. Uluski. 2003. Plan Ahead for Substation Automation. *IEEE Power & Energy Magazine*, 1(2):32-41.

Hardison, J. and R. Martin. 2000. Plant SCADA System Enables Optimization Initiative. In *Proc. 2000 AWWA Annual Conference and Exposition, June 11-15, 2000, Denver, Colorado.* Denver, Colo.: AWWA.

Hargesheimer, E., O. Conio and J. Popovicova. 2002. *Online Monitoring for Drinking Water Utilities.* Denver, Colo.: AWWARF, AWWA and CRS PROAQUA.

Hill, R. 1997. *Automated Process Control Strategies.* Water Environment Federation. Alexandria, VA.

Hill, R., B. Manross, and A. Manning. 2001. Assess Installed State-of-the-Art WWTP Sensing and Control Systems. In *Proc. Of Water Environment Federation Annual Technical Conference & Exposition, October 13-17, 2001, Atlanta, Ga*. Alexandria, VA: WEF.

Hill, R.D., R.C. Manross, E.V. Davidson, T.M. Palmer, M.C. Ross, and S.G. Nutt. 2002. *Sensing and Control Systems: A Review of Municipal and Industrial Experiences*. Alexandria, Va: WERF

Hinthorn, R., F. Moshavegh, L. Wilson, S. Yadav. 2003. A Vision for Real-time Monitoring and Modeling of Water Quality in Water Distribution Systems. In *Proc. Of AWWA Information Management Technology Conference, April 27-30, 2003, Santa Clara, CA*. Denver, Colo.: AWWA.

Hoffman, M.R. and L.K. Reynolds. 1994. The Role of Control Systems in the Management of a Modern Water Utility. In *Proc. Of IEEE Conference on Control Applications, August 24-26, 1994, Glasgow, UK*. New York, N.Y.: IEEE.

Huntington, R. 1998. Twenty Years Development of ICA in a Water Utility. *Water Science and Technology*, 37(12):27-34.

IEEE (Institute of Electrical and Electronics Engineers). 1996. Guide for Computer-Based Control for Hydroelectric Power Plant Automation. IEEE Std 1249-1996. New York, N.Y.: IEEE. Approved December 10, 1996.

Illinois Administrative Code. Title 35, Subtitle F, Public Water Supplies, Chapter I, Pollution Control Board, Part 603, Ownership and Responsible Personnel, and Part 611, Primary Drinking Water Standards, Subparts L, M, N, O, Q. http://www.ipcb.state.il.us/SLR/IPCBandIEPAEnvironmentalRegulations-Title35.asp

Jacobs, J., T. Kerestes and W.F. Riddle. 2003. *Best Practices for Energy Management*. Denver, Colo.: AwwaRF.

Jentgen, L., S. Conrad, and T. Lee. 2005. Optimizing System Operations. *Jour. AWWA*, 97(8): 58-66.

Johnson, D. 2004. Beyond Process Control. *Water and Wastewater Products*, 4(6).

Junnier, R.J. 1997. Developing Cost/Benefits for the Implementation of Computer Technology ay Water & Wastewater Utilities. In *Proc. Of Computer Technologies for the Competitive Utility, Water Environment Federation, July 15-18, 1997, Philadelphia, Pennsylvania. Alexandria*, Va.: WEF.

Kaplan, R.S., and D.P. Norton. 1996. *The Balanced Scorecard*. Boston, MA: Harvard Business School Press.

Keen J.M., B. Digrius. 2003. *Making Technology Investments Profitable*. Hoboken, New Jersey: John Wiley & Sons.

Kendricks, L.E., Jr. 1999. Reaping the Benefits of Wastewater Treatment Plant Automation. *Pollution Engineering*. 31(12):52-53.

Kingham, T.J. and T.M. Hoggart. 1995. Chlorination Control in a Large Water Treatment Works. In IEE Colloquium on Application of Advanced PLC Systems with Specific Experiences from Water Treatment (Digest No. 1995/112), June 29, 1995, London, UK. IEE.

Kugelman, I. and J. Houthoofd. 2002. *Optimization of Treatment Plant Operation*. EPA/600/J-85/218 (NTIS PB86118486). Washington, D.C.:USEPA.

LaMontagne, P.L. Degrees of Automation. 2004. *Water Environment & Technology*,16(11):43-45.

Lant, P. and M. Steffens. 1998. Benchmarking for Process Control: "Should I Invest in Improved Process Control?". *Water Science and Technology*. 37(12):49-54.

Latimer, E.A., and D.L. Reddell. 1990. Components for an Advanced Rate Feedback Irrigation System (ARFIS). *Transactions of the American Society of Agricultural Engineers*. 33(4):1162-1170.

Lauer, W.C. 2005. Automation Supports Unattended Operation. *Opflow*, 31(2):7.

Lehtonen, M. and S. Kupari. 1995. A Method for Cost Benefit analysis of Distribution Automation. In *Proc. of EMPD '95 International Conference on Energy Management and Power Delivery, November 21-23, 1995, 1:49-54*. New York, N.Y.: IEEE.

Lott, C. 2006. The Investment FAQ. [online] Available: http://invest-faq.com

Marlin, T.E., M.L. Brisk, G.W. Barton, and J.D. Perkins. 2001. Experiences from an Industry-University Study on Control Benefits. *Pulp & Paper Canada*, 102(4):34-37.

Maskey, R., G. Roberts, and B. Graetz. 2001. Farmers' Attitudes To The Benefits And Barriers Of Adopting Automation For Surface Irrigation On Dairy Farms In Australia. *Irrigation and Drainage Systems*. 15:39-51.

Maxwell, S. 2005. An Overview of Trends in the Drinking Water and Wastewater Treatment Markets. *Jour. AWWA*, 97(9): 4-5.

McIntyre, J. 2003. *Water Treatment Regulation in the United States and Effects on the Global Innovation Process*. Working Paper. Georgia Institute of Technology, Atlanta, Ga.

Means III, E.G. 2004. *Water and Wastewater Industry Energy Efficiency: A Research Roadmap*. Denver, Colo.: AwwaRF and AWWA.

Means III, E.G., J. Bernosky, and R. Patrick. 2006. Technology Trends and Their Implications For Water Utilities. *Jour. AWWA*, 98(1): 60-71.

Means III, E.G., L. Ospina, and R. Patrick. 2005. Ten Primary Trends and Their Implications for Water Utilities. *Jour. AWWA*, 97(7): 64-77.

Michigan Department of Environmental Quality, Water Division, Supplying Water to the Public, Part 7; R325.10701-R325.10738, Surveillance, Inspection and Monitoring and Part 9 R325.11901-R325.11918, Examination and Certification of Operators. http://www.michigan.gov/deq/0,1607,7-135-3307_4132-14902--,00.html

Morris, J.F., F.J. Kern, and E.F. Richards. 1988. Distribution Automation of the Association of Missouri Electric Cooperatives – A Statewide Evaluation of Load Management. *IEEE Transactions on Industry Applications*, 24(5):782-791, September/October 1988.

National Drinking Water Clearinghouse. 1997. *Package Plants*. Tech Brief, November 1997, 6:1-4.

New York State Department of Health, Title 10, Part 5, Subpart 5-1, Public Water Systems. http://www.health.state.ny.us/nysdoh/water/part5/subpart5.htm

Newton, C. 1995. Benefits Analysis for Automation Programs. *Transmission & Distribution*. 47(7):12.

O'Brien, A., Swanback, S., and Marks, K. 2000. Visioning a New California Wastewater Treatment Plant Unattended Operation/Unrestricted Effluent Use. 2000. In *Proc. of Water Environment Federation Annual Technical Conference & Exposition, October 14-18, 2000, Anaheim, CA*. Alexandria, VA: WEF.

Ohio Administrative Code, Part 3745-81, Primary Drinking Water Rules and Part 3745-7-04 (Draft), Treatment works and sewerage system classification and staffing requirements. http://www.epa.state.oh.us/ddagw/oac.html#374507

Ohto, T. 1998. Controls, Computers and Communications: Fusion in Instrumentation, Control and Automation of Water and Wastewater Systems in Japan. *Water Science and Technology*. 37(12):15-19.

Olsson, G. and P. Ingildsen. 2003. Automation in Wastewater Treatment Plants. In *Proc. of Water Environment Federation Annual Technical Conference & Exposition, October 12-15, 2003, Los Angeles, CA.* Alexandria, VA: WEF.

Olsson, G., M. Nielsen, Z. Yuan, A. Lynggaard-Jensen, J-P. Steyer. 2005. *Instrumentation, Control and Automation in Wastewater Systems*. London: IWA Publishing.

Opincar, V. 2003. Automated Surveillance for Water Utilities. *Jour. AWWA*, 95(9):48-51.

Pacific Gas & Electric. 2006. Industrial/General Service E-20 Electric Rate Schedule. [online] Available: http://www.pge.com/rates/tariffs/IndustrialCurrent.xls

Patrick, R., J. Rompala, A. Symkowski, W. Kingdom, R. Serpente, N. Freeman, B. Stevens, C. Koch, and T. Kochaba. 1997. *Benchmarking Wastewater Operations – Collection, Treatment, and Biosolids Management*. Alexandria, Va.: WERF.

Pennsylvania Code, Title 25, Environmental Protection, Chapter 109, Safe Drinking Water, Subchapter C, Monitoring Requirements and Subchapter G, System Management Responsibilities. http://www.pacode.com/secure/data/025/025toc.html

Pisello, T. 2001. *Return on Investment for Information Technology Providers*. New Canaan, Conneticut: Information Economics Press.

Porter, Rolfe. 1996. Plant Energy Audits and Pumping Efficiency Optimization. In *Proc. AWWA Annual Conference and Exposition, June 26, 1996, Toronto, Canada.* Denver, Colo: AWWA.

Pramanik, A., P. LaMontagne, and P. Brady. 2002. Automatic Improvements. *Water Environment & Technology*. 14(10):46-50.

Preble, C. and T. Valorose. 1995. Computer at Water Plant Runs Second Plant by Remote Control. *Water Engineering & Management*, 142(5):34-38.

Quick, G., R. Emanuel, J. O'Connor. 1997. City of Austin's Control System Modernization Yields Operational Benefits for the South Austin Regional Wastewater Treatment Plant. In *Proc. of Water Environment Federation Annual Technical Conference & Exposition, October 18-22, 1997, Chicago, IL.* Alexandria, VA: WEF.

Rahimi, A.F., L.P. Hajdu, L. Kiss, and L. Balogh. 1993. A General Cost-Benefit Analysis Methodology for the Evaluation of EMS/SCADA Procurement Alternatives. In *Proc. IEEE/NTUA Athens Power Tech Conference, September 5-8, 1993, Athens, Greece.* New York, N.Y.: IEEE.

Rau, M. 1990. Building Momentum for Automatic Meter Reading. *Water Engineering & Management*, 137(3):32-33.

Reely, B.T., L.W. House. 2004. *Demand Response Engineering Analysis of El Dorado Irrigation District's El Dorado Sub-System.* Efficiency Analysts International. http://www.waterandenergyconsulting.com/EIDfinal.pdf

Reis, J., l. Loy, R. Metzger. 1999. *Water Treatment Plant Filter Backwash Optimization Study*. Final Report 00-08, November 1999. Albany, Ny, NYSERDA

Renner, R., B. Hegg, and J. Bender. 1990. *Optimizing Water Treatment Plant Performance With the Composite Correction Program*. EPA/625/8-90/017. Washington D.C. USEPA.

Rice, M.D., R. Edmonson, and L.R. Smith. 2005. Single Source Responsibility = Seamless SCADA Expansion and Integration. In *Proc. Of AWWA Annual Conference and Exposition, June 12-16, 2005, San Francisco, CA.* Denver, Colo.: AWWA.

Roberts, G.V. 1999. Analysis of Reliability of Networks and Justification of Automation. In *IEE Colloquium on Remote Control and Automation on 11kV Networks Beyond the Primary Substation (Ref. No. 1999/195), November 22, 1999*. IEE.

Ross, B., B. Brunner, L. Mincy, and G. Jones. 2003. Reinventing SCADA in the 21st Century. In *Proc. Of Water Environment Federation Annual Technical Conference & Exposition, October 12-15, 2003, Los Angeles, CA*. Alexandria, VA: WEF.

Russo, P. 2001. District Wide Conversion to Unmanned Weekends and Nights. In *Proc. Of Water Environment Federation Annual Technical Conference & Exposition, October 13-17, 2001, Atlanta, Ga.* Alexandria, VA: WEF.

Schlenger, D.L., and W.F. Riddle. 1993. Can Computers Replace People in Water Treatment Operations? Guidelines for Staffing Adjustments Through Automation. In *Proc. 1993 AWWA Annual Conference, June 6-10, 1993, San Antonio, Texas*. Denver, Colo.: AWWA.

Schlenger, D.L., W.F. Riddle, B.K. Luck, and M.H. Winter. 1996. *Automation Management Strategies for Water Treatment Facilities*. Denver, Colo.: AwwaRF and AWWA.

Schoedinger, S. and D. Diffee. 1999. Water/Wastewater Treatment Plant Automation Strategies for Competitiveness and Efficiency. In *Proc. Florida Section AWWA Annual Conference, November 29 – December 2, 1999, Orlando, FL*. Denver, Colo.: AWWA.

Schultz, J. 2003. Electric Energy Usage Hemlock Water Treatment Plant Case Study. New Haven Energy Workshop for Water & Wastewater Facilities – Clean Water, Clear Skies, and Cost Savings: What Drinking Water and Wastewater Utilities Can Learn From Energy Efficiency Pioneers in Connecticut, November 20, 2003. http://www.summersaver.state.ct.us/documents.htm

Shariff, R. A. Cudrak, Q. Zhang, and S. Stanley. 2003. Integration of Next Generation Process Control Strategies into Existing Water Treatment Plants. In *Proc. Of AWWA Information Management Technology Conference, April 27-30, 2003, Santa Clara, CA*. Denver, Colo.: AWWA.

Shariff, R., A. Cudrak, E. Saumer, and S. Stanley. 2003. Improve Performance and Decision-Making Through Greater Utilization of Plant Data. In *Proc. Of AWWA Information Management Technology Conference, April 27-30, 2003, Santa Clara, CA*. Denver, Colo.: AWWA.

Shariff, R., A. Cudrak, W. Stachowski, R. Welz, and S. Stanley. 2002. Centralized Remote Operation of Multiple Water and Wastewater Plants. In *Proc. Of AWWA Information Management Technology Conference, April 14-17, 2002, Kansas City, MO*. Denver, Colo.: AWWA.

Shariff, R., R. Welz, R., S. Stanley, W. Stachowski, and R. Corscadden. 2001a. Automation and Unattended Operation of Large Water Plants. In *Proc. Of AWWA Information Management Technology Conference, April 8-11, 2001, Atlanta, Georgia*. Denver, Colo.: AWWA.

Shariff, R., R. Welz, S. Stanley, W. Stachowski, and R. Corscadden. 2001b. Remote Monitoring and Operation of Isolated Facilities in Cold Regions. In *Proc. of AWWA Annual Conference and Exposition, June 17-20, 2001, Washington D.C.* Denver, Colo.: AWWA.

Singh, A. 2001. An Economic Feasibility Analysis for the Automation of Hazardous Waste Landfills. *Cost Engineering*, 35(4):13-21.

Smith, H., R. Emanuel, B. Jacobsen and M. Hartson. 2000. Integrating Maintenance and Operations: The Critical Information Link for Unattended Operations. In *Proc. 2000 AWWA Distribution System Symposium, September 10-13, 2000, New Orleans, Louisiana*. Denver, Colo.: AWWA.

Smith, H., R. Emanuel, B. Jacobsen, and W.L. Overbeek. Automation Strategies for Unattended Water Treatment Plant Operation; OUC's Water Project 2000 Experience. In *Proc. 1999 AWWA Information Management & Technology Conference, April 18-21, 1999, New Orleans, Louisiana*. Denver, Colo.: AWWA.

Smith, H., R. Emanuel, M. Wehmeyer and B. Phillips. 1997. Alternative Project Delivery for Facility Automation and Information Management for a Multi-Plant, Systemwide Infrastructure Modernization. In *Proc. 1997 AWWA Computer Conference, April 13-16, 1997, Austin, Texas.* Denver, Colo.: AWWA.

Terry, W.W. 2002. Hydro Automation Program Improves Efficiency and Reduces Operating Expense at TVA. *Power Engineering*. 106(3):54-60.

Texas Administrative Code, Title 30, Environmental Quality, Part 1, Chapter 290, Public Drinking Water, Subchapter D, Rules and Regulations for Public Water Systems, and Subchapter F, Drinking Water Standards Governing Drinking Water Quality and Reporting Requirements for Public Water Systems. http://info.sos.state.tx.us/pls/pub/readtac$ext.ViewTAC?tac_view=4&ti=30&pt=1&ch=290

Thomas, L., J. Kalinowski, C.Cook, and L. Verduzco. 2004. Fieldbus Instruments Support a Large Vision. *InTech*, July 2004. http://www.isa.org/InTechTemplate.cfm?Section=Article_Index1&template=/ContentManagement/ContentDisplay.cfm&ContentID=37013

U.S. Department of Commerce, Office of the Assistant Secretary for Conservation and Renewable Energy, Federal Energy Management Program, NIST Handbook 135, 1995 Edition. Life Cycle Costing Manual for the Federal Energy Management Program. http://www.bfrl.nist.gov/oae/publications/handbooks/135.pdf

U.S. Department of Energy, Office of Energy Efficiency and Renewable Energy, Federal Energy Management Program. 1996. 10 CFR 436, Subpart A, Methodology and Procedures for Life Cycle Cost Analyses. http://www.access.gpo.gov/nara/cfr/waisidx_04/10cfr436_04.html

U.S. Environmental Protection Agency. 2000. *Community Water System Survey 2000*. EPA-815-R-02-005A. http://www.epa.gov/safewater/cwssvr.html

U.S. Environmental Protection Agency. 2001. 1999 *Drinking Water Needs Survey, Modeling the Cost of Infrastructure*. EPA 816-R-01-005. http://www.epa.gov/ogwdw/needs/fullcost.pdf

U.S. Environmental Protection Agency. 2003. *Small Drinking Water Systems Handbook: A Guide to "Packaged" Filtration and Disinfection Technologies with Remote Monitoring and Control Tools, May 2003*. EPA-600-R-03-041. http://www.epa.gov/ord/NRMRL/Pubs/600R03041/600R03041.pdf

U.S. Environmental Protection Agency. 2005. 40 CFR Parts 141, 142, and 143. National Primary and Secondary Drinking Water Regulations, 336-615. http://www.access.gpo.gov/nara/cfr/waisidx_02/40cfrv19_02.html

U.S. Environmental Protection Agency. 2005. *Drinking Water Infrastructure Needs Survey and Assessment, Third Report to Congress*. EPA 816-R-05-001. http://www.epa.gov/ogwdw000/needssurvey/pdfs/2003/report_needssurvey_2003.pdf

U.S. Federal Government, Office of Management and Budget. 1992. Memorandum for Heads of Executive Departments and Establishments. Guidelines and Discount Rates for Benefit-Cost Analysis of Federal Programs. Circular No. A-94 Revised (Transmittal Memo No. 64). http://www.whitehouse.gov/omb/circulars/a094/a094.html.

U.S. General Accounting Office. 2004. Water Infrastructure Comprehensive Asset Management Has Potential to Help Utilities Better Identify Needs and Plan Future Investments. GAO-04-461. Washington, D.C.: GAO. http://www.gao.gov/new.items/d04461.pdf

Wensloff, David A. 1998. Optimizing Your Industrial Wastestream Costs. 1998. *Water/Engineering & Management*, 145(3):26.

White, G.C. 1999. *Handbook of Chlorination and Alternative Disinfectants*. 4th ed. New York: John Wiley & Sons, 1999.

Wu, M.D. and J.C. Liu. 1999. Fuzzy Control of a Coagulation Reaction for the Treatment of High-Turbidity Water. *Aqua (Journal of Water Services Research and Technology)*. 48(5):211-217.

Younkin, C. and G. Huntley. Unattended Facilities Offer Competitive Advantage. In *Proc. 1996 AWWA Computer Conference, April 21-24, 1996, Chicago, Illinois*. Denver, Colo.: AWWA.

ABBREVIATIONS

AWWA	American Water Works Association
ASCE	American Society of Civil Engineers
AwwaRF	Awwa Research Foundation
CEC	California Energy Commission
CWS	Community Water Systems
DCU	Distributed Control Unit
EPRI	Electric Power Research Institute
FIT	Flow Indicating Transmitter
GAO	Government Accountability Office
GIS	Geaographic Information System
HMI	Human Machine Interface
I/O	Input/Output
ISA	Instrument Society of America
KW	Kilowatt
KWh	Kilowatt Hour
LIMS	Laboratory Information Management Systems
LIT	Level Indicating Transmitter
Mgd	Million Gallons per Day
MTBF	Mean Time Before Failure
NPV	Net Present Value
O&M	Operations and Maintenance
P&ID	Process and Instrumentation Diagram
PLC	Programmable Logic Controller
ROI	Return On Investment
RTU	Remote Terminal Unit

SCADA	Supervisory Control And Data Acquisition
SCD	Streaming Current Detector
UPS	Uniterruptible Power Supply
USEPA	United States Envirnonmental Protection Agency
VFD	Variable Frequency Drive
WTP	Water Treatment Plant

IWA Publishing's authorised EU representative for General Product Safety Regulations is Diane D'Arras, 15 rue Duret, 75116 Paris, France, e-mail: safety@iwap.co.uk.

Printed and bound by CPI Group (UK) Ltd, Croydon, CR0 4YY
17/09/2026
02218808-0004